New Theories for Chemistry

New Theories for Chemistry

Jan C.A. Boeyens
Department of Chemistry
University of Pretoria
Pretoria
South Africa

2005

ELSEVIER

Amsterdam • Boston • Heidelberg • London • New York • Oxford • Paris
San Diego • San Francisco • Singapore • Sydney • Tokyo

ELSEVIER B.V.
Radarweg 29
P.O. Box 211, 1000 AE Amsterdam
The Netherlands

ELSEVIER Inc.
525 B Street, Suite 1900
San Diego, CA 92101-4495
USA

ELSEVIER Ltd
The Boulevard, Langford Lane
Kidlington, Oxford OX5 1GB
UK

ELSEVIER Ltd
84 Theobalds Road
London WC1X 8RR
UK

First edition 2005

Library of Congress Cataloging in Publication Data
A catalog record is available from the Library of Congress.

British Library Cataloguing in Publication Data
A catalogue record is available from the British Library.

ISBN: 0 444 51867 3

⊛ The paper used in this publication meets the requirements of ANSI/NISO Z39.48-1992 (Permanence of Paper).
Printed in The Netherlands.

Preface

The progress of theoretical chemistry during the 20th century was severely hampered by a misreading of quantum theory and relativity. Since it was first pointed out by Dirac [1] that quantum theory, correctly applied, reduced chemistry to a subset of physics, the quest to find the quantum operators that generate the observables of chemistry was on. This quest has been singularly sterile and the reasons for the failure are becoming clear only now. The concepts of chemistry reflect the behaviour of electrons, atoms and molecules in three-dimensional space. On the other hand, theoretical studies sought to generate these concepts by operating on multidimensional real wave functions in configuration space.

The debate on how to interpret quantum theory raged for decades without addressing the fundamental problems of chemistry. With the nature of the elementary quantum entities at issue, the bigger question about molecular identity remained unanswered. While the particle concept is being confused with probability density and matter waves there is little prospect of an operational definition of the important objects of chemistry. The real dilemma has been how to reconcile the consistent classical theories of chemistry with the concepts of quantum mechanics. One option was to formulate theoretical chemistry in terms of classical and non-classical variables by a non-commutative algebra of observables that makes provision for the two sectors [2]. In this scheme all variables pertaining to molecular conformation are identified as classical, although the molecular electron-density function is clearly non-classical. Concepts such as electronegativity with a non-classical ring to it, but without an operational quantum definition, had to be banned from the vocabulary.

A second debate about the completeness of quantum theory did not benefit theoretical chemistry any better. Superposition of state functions which is allowed in quantum, but not in classical systems, dictates that the former is an entangled, non-local holistic theory [3]. The famous Einstein-Bohr debates, although centred around this issue, became so bogged down in side issues that they never squarely faced the real dilemma that a non-local (quan-

v

tum) theory contradicts the special theory of relativity. The amazing fact that sterically unlikely molecular rearrangements can only be comprehended in terms of non-local theory of molecular structure, was simply ignored. In support of Bohr's position Von Neumann provided mathematical proof that no concealed parameters (hidden variables) could be introduced to transform the indeterministic description provided by quantum theory into a deterministic one, as advocated by Einstein.

> But in 1952 I saw the impossible done. It was in the papers of David Bohm [4].

This quotation is from J.S. Bell [5]. The event referred to, marks the beginning of a slow, but certain return to the original spirit of quantum theory to understand the stability and structure of ponderable matter. Bohm's ideas have been resisted by the physics community but, as pointed out by Bell,

> even Pauli, Rosenfeld and Heisenberg, could produce no more devastating criticism of Bohm's version than to brand it as 'metaphysical' and 'ideological'.

Einstein, who made an unsuccessful attempt in 1927 with the same objective as Bohm [6] and whom Bohm acknowledged for several interesting and stimulating discussions in his paper, described[1] Bohm's 1952 work as "too cheap".

The chemistry community, understandably, failed to respond at all, even though Bohmian mechanics probably holds the key to the development of a theory of chemistry, soundly based on quantum theory and relativity. The problem with molecular structure is resolved by the ontological interpretation of quantum-mechanical orbital angular momentum and the key to chemical reactivity and reaction mechanism is provided by the quantum potential function.

The theories of quantum mechanics and relativity and the concepts underlying these theories are unfamiliar territory for many chemists. To prepare the ground for a reassessment of the chemical importance of these theories against the background of Bohmian mechanics, the relevant concepts have recently been presented from a chemical perspective [7]. Constant reference to this earlier work will be made here. Important equations, discussed and

---

[1]Between submission and publication of Bohm's paper he was forced out of Princeton and the USA by the House Un-American Activities Committee.

numbered in [7] will be referred to in this work by a T prefixed to their numbers.

The new theories that spring from the application of Bohmian mechanics to chemical problems reveal a close connection between chemical phenomena and the attributes of space-time. The most fundamental principle of chemistry is the periodic classification of the elements in terms of natural numbers. Examined against the background of number theory a deeper level of periodicity that embraces all nuclides is revealed and found to relate on a cosmic scale to an involution in space-time structure.

Virtually no aspect of chemistry is left untouched by the new insight gained through consideration of Bohmian mechanics, number theory and symmetry. A preliminary analysis of some of these aspects are presented here in an effort to stimulate further research, that may happen once familiarity with the new approach overcomes traditional scepticism. Most of the material in this book has been published in somewhat different form before. Chapters 1, 2, 4, 5, and 6 are elaborations based on recent review articles [8, 9, 10, 11] by the author and included here with permission by the relevant publishers.

Awareness that the laws of Nature are firmly based on the symmetry and topology of space-time developed over a period of almost two decades. I am indebted to former colleagues and students at the University of the Witwatersrand, Johannesburg for many hours of discussion, speculation and argument around this issue. Foremost among these are Demi Levendis, Spike Mc Carthy, Charles Marais, Johan Prins and Don Travlos. My attempts to formulate some of the wild ideas into comprehensible form continued in Europe during two periods of sabbatical leave at the Free University, Berlin and the Ruprecht-Karls University, Heidelberg, Germany, made possible by the award of an Alexander von Humboldt Prize. The writing was completed during my tenure at the University of Pretoria. I gratefully acknowledge stimulating interaction with Anton Amann, Peter Comba, Danita de Waal, Werner Gans, Tibor Koritsanszky, Peter Luger, John Ogilvie and Casper Schutte, whom I implicate in no way as coresponsible for any contentious statements. Finally, my appreciation to members of my immediate family whom I fear, had to put up many times with my excursions into another dimension.

Pretoria
November 2004

Contents

Chapter 1

The Symmetry Laws of Nature

1.1 Introduction

When first experienced science presents itself as a bewildering collection of
unrelated observations and theories, but closer examination almost invariably
reveals unexpected common ground. The broader the view, the more uniform
the landscape becomes. It is therefore not surprising that, for centuries,
philosophers have been looking for and finding, the ultimate basis of science.
The modern era is no exception in this respect. A variety of concepts, ranging
from quantum theory and relativity to chemical interaction, biology and
behavioural patterns, are found to share a few common rudiments, of which
symmetry is one of the more conspicuous. The need to quantify this unifying
concept has inevitably stimulated the growth of appropriate mathematical
structures, to the point where it may now be feasible to follow the symmetry
thread through all of science, and certainly through most of the physical
sciences.

Ideas originating from a study of electromagnetism, mechanics, thermodynamics and other diverse fields may well appear to be unrelated until they
are seen to be based on common laws dictated by the vacuum structure and
the symmetry of space-time. Proper understanding of this implied synthesis
of ideas, requires familiarity with the concepts of group theory, fields, waves,
particulate matter, relativity, quantum mechanics, phase relationships and
solid-state theory. It may be rewarding to have a new look at science as
a unified pursuit, against this background. The first objective will be to
get a clear understanding of the symmetry concept itself and to eliminate
any confusion that may arise from its use in different contexts. There are
symmetrical objects and operations, internal symmetry and symmetry in the
laws of physics. These different symmetries may be discrete or continuous

and be of either finite or infinite order.

1.2   Group Theory and Symmetry

Although group theory formally is a branch of abstract mathematics its development has been linked intimately with concepts and operations related
to the symmetry of physical objects and structures.

   The concept of symmetry is of ancient vintage and in many ways almost identical with the equally elusive concepts of beauty and harmony,
i.e. beauty of form arising from balanced proportions. Although symmetry
can be described in mathematically precise terms, symmetry in the physical
world, like beauty[1], never absolutely obeys the mathematical requirements
of group theory: even the most perfect crystal has a surface that spoils the
symmetry.

   It will be argued that any interaction represents a broken symmetry and
that mathematically precise symmetry never occurs in the real world. The
implication is that the almost magical parallel between mathematics and
physics that fascinates both scientists and philosophers, breaks down at the
most fundamental level. This observation has little or no impact on the
analysis of physical systems (*e.g.* in crystallography) where exact symmetry
is no more than a simplifying model. However, the common dogma of time-
reversible laws of physics, may in the view of all this, introduce a fallacy that
re-emerges as the paradox of irreversability only much later.

   In the first instance however, symmetry will be treated in what follows as
precisely described by the formalism of group theory and any change in symmetry of a system will be viewed as a transition between perfect symmetry
types.

   Most of the concepts of group theory can readily be demonstrated in terms
of operations on a pointer that moves over the dial of a clock *i.e.* translations
on the unit circle. The one-dimensional symmetry that relates operations on
an elementary dial with four allowed equivalent positions E, A, B, C among
which the pointer may move by discontinuous clicks, represents a group with
four elements. The same symbols that identify the fixed points on the dial
may also serve to represent the group operations corresponding to clockwise
rotations of $n\pi/2$ for $n = 1(A), 2(B), 3(C), 4(E)$. The composition (product)
of any two operations corresponds to successive rotations according to the
following multiplication table:

---

[1]It is the beauty of a theory which is the real reason for believing in it - Dirac [12]

Figure 1.1: *Symmetry-related positions on the dial of a clock.*

G	E	A	B	C
E	E	A	B	C
A	A	B	C	E
B	B	C	E	A
C	C	E	A	B

The operations E, B are seen to constitute a *subgroup* of G with elements $n\pi/1$, $n = 1, 2$:

	E	B
E	E	B
B	B	E

In general, it follows that the set of allowed rotations $n\pi/m$ for integral m and $n = 1, 2, ..., 2m$ constitutes a group of order $2m$, with a subgroup of order $2m/p$ for each integral $p = 2m/n > 1$. The elements of a given subgroup are the rotations $n\pi/m$ for $np = 1, 2, ..., 2m$. The group generated by $m = 3$ is readily shown to be of order 6, with subgroups of order 3, 2 and 1.

The equivalent positions on the face of a digital clock that only registers hours, correspond to twelve (2m) rotations with subgroups of order 1,2,3,4 and 6, and elements (12), (12,6), (12,4,8), (12,3,6,9), (12,2,4,6,8,10). A digital minute clock represents a group of order 60 with subgroups of order 1,2,3,4,5,6,10,12,15,20,30 - *i.e.* the factors of 60. The dial of any digital clock, however small the unit interval may be, represents a group of the same type, but of higher order. A clock with analogue motion and a pointer that moves continuously along the dial, represents a one-parameter infinite continuous group, *i.e.* the Lie group O(2).

The hands of analogue and digital clocks rotate in two fundamentally different ways, characterized by continuous and discrete symmetry groups respectively. The continuous rotation during a complete cycle of 2π is isomorphic with translations on the real line and has an infinite number of equivalent positions. The moving pointer touches all of these points during one rotational cycle. In digital mode the pointer also visits all equivalent points, which could also be infinite in number.

The points on the dial are said to be related by rotational symmetry, referred to an axis of order n, corresponding to the number of points allowed on the circumference of the circle of rotation. The various dials considered above, have rotational symmetry axes of order 4, 6, 12, 60 and ∞. The allowed rotations in each case constitute a group of which only the last may be continuous.

There is another way to interpret the progress of a hand that moves across the dial of a clock, when it measures time. Each position of the hand denotes a specific time, which is different from all others, and only repeats itself after a complete cycle of either 2π or 4π, depending on the definition of a unit cycle as either 12 or 24 hours. The symmetry that describes the progress of the moving hand is equivalent to that of translational motion over a potential field that repeats at periodic intervals, for instance in a crystal.

The known symmetries that occur in the physical world can be described in terms of symmetry groups which in principle are all of the same type as those represented by either the analogue or digital clock, also in higher-dimensional space. The most commonly encountered symmetries are discussed below.

1.2.1   Continuous Space-Time Symmetries

(i) Translations in space, $x \rightarrow x + a$, where a is a constant 3-vector: This symmetry, applicable to all isolated systems, is based on the assumption of *homogeneity of space*, *i.e.* every region of space is equivalent to every other, or alternatively, physical phenomena must be reproducible from one location to another. It will be shown that the conservation of linear momentum is a consequence of this symmetry.

(ii) Translations in time, $t \rightarrow t + a_0$, where a_0 is a constant: This symmetry, applicable also to isolated systems, is a statement of *homogeneity of time*, *i.e.* given the same initial conditions, the behaviour of a physical system is independent of the absolute time - in other words, physical phenomena are reproducible at different times. The conservation of energy will be shown to derive from this symmetry.

(iii) Rotations in three-dimensional space, $x_i \rightarrow x_i' = R_{ij}x_j$, where $i, j = 1, 2, 3$, $\{x_i\}$ are the three components of a vector and R is a 3×3 (orthogonal) matrix: This symmetry reflects the *isotropy* of space, *i.e.* the behaviour of isolated systems must be independent of the orientation of the system in space. It will be shown to lead to the conservation of angular momentum.

(iv) Lorentz transformations

$$\begin{pmatrix} t \\ \boldsymbol{x} \end{pmatrix} \rightarrow \begin{pmatrix} t' \\ \boldsymbol{x'} \end{pmatrix} = \Lambda \begin{pmatrix} t \\ \boldsymbol{x} \end{pmatrix}$$

where Λ is a 4×4 Lorentz matrix and $\boldsymbol{x}$ stands for a three-component column vector: This symmetry embodies the generalization of the classical, separate space and time symmetries into a single *space-time symmetry*, known as *special relativity*. The conservation of energy-momentum is a consequence of this symmetry.

1.2.2 Discrete Space-Time Symmetries

1. Space inversion (or parity transformation), $\boldsymbol{x} \rightarrow -\boldsymbol{x}$: This symmetry is equivalent to the reflection in a plane (*i.e.* mirror symmetry), as one can be obtained from the other by combination with rotation through angle π.

2. Time reversal transformation, $t \rightarrow -t$: This is like space inversion and most likely space-time inversion is a single symmetry that reflects the local euclidean topology of space, observed as the conservation of matter.

3. Discrete Rotational Symmetry: This is a subset of continuous rotations and reflections in three-dimensional space. Since rotation has no translational components their symmetry groups are known as *point groups*. Point groups are used to specify the symmetry of isolated objects such as molecules.

4. Discrete translations on a lattice. A periodic lattice structure allows all possible translations to be understood as ending in a confined space known as the *unit cell*, exemplified in one dimension by the clock dial. In order to generate a three-dimensional lattice, parallel displacements of the unit cell in three dimensions must generate a space-filling object, commonly known as a crystal. To ensure that an arbitrary displacement starts and ends in the same unit cell it is necessary to identify opposite points in the surface of the cell. A general translation through the surface then re-enters the unit cell from the opposite side.

The eight corners of a unit cell shaped like a parallepiped are identical because of lattice, or translational symmetry along its edges, called the *crystallographic axes*. The lattice symmetry is described by a *space group* and the resultant of any displacement can be decomposed into three components

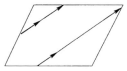

Figure 1.2: *Translation through a crystal lattice mapped into a single unit cell.*

known as the *fractional coordinates.* Lattices of higher symmetry may contain elements of point-group symmetry, in addition to the identity operation, giving a total of 230 possible space groups. Any space group of high symmetry contains one or more space groups of lower symmetry as subgroups. The space group of lowest symmetry ($P1$) is a subgroup of all other space groups.

1.2.3 Permutation Symmetry

Systems containing more than one identical particles are invariant under the interchange of these particles. The permutations form a symmetry group. If these particles have several degrees of freedom, the group theoretical analysis is essential to extract symmetry properties of the permissible physical states. Examples include Bose-Einstein, Fermi-Dirac, Maxwell-Boltzmann statistics, Pauli exclusion principle, *etc.*

1.2.4 Gauge Invariance and Internal Symmetry

The idea of gauge invariance is best understood in terms of a complex phase transformation,

$$\psi(x) \to e^{i\alpha}\psi(x)$$

where α is a real constant. This transformation does not involve any space-time parameters and represents an *internal symmetry.* The family of phase transformations $U(\alpha) \equiv e^{i\alpha}$, where a single parameter α may run over real numbers, forms a unitary Abelian group known as the $U(1)$ group. It has been shown to be related to the conservation of charge. Other internal symmetries of nuclear and elementary particle physics relate to the conservation of isospin, colour, *etc.*

The $U(1)$ gauge transformation of the form

$$\psi \to \psi' = \psi e^{i\alpha} \tag{1.1}$$

is called a global phase transformation since α is independent of space and time. Should one introduce a space-time dependence into the parameters, the corresponding gauge transformation is said to be local. Taking a derivative of ψ then generates an extra term and to compensate for this an additional field with transformation

$$A_\mu \rightarrow A'_\mu = A_\mu + \partial_\mu \alpha$$

must be introduced. The Maxwell field is the best known example of such an operation. Alternatively it is possible to start from (1) and work out the details of the Maxwell field.

1.3 Symmetry and the Laws of Nature

The so-called laws of Nature are scientific generalizations of regularities observed in the behaviour of a system under specified conditions. Behaviour in this sense implies, almost invariably, the way in which a system of interest develops as a function of time. More basic still, more than law, call it axiom, is the all but universally accepted premise that the outcome of any scientific experiment is independent of its location and orientation in three-dimensional space, provided the experimental conditions can be replicated. A moment's reflection shows that this stipulation defines a symmetry which is equivalent to the conviction that space is both *homogeneous* and *isotropic*. The surprising conclusion is that this reproducibility, which must be assumed to enable meaningful experimentation, dictates the nature of possible observations and hence the laws that can be inferred from these observations. The conclusion is father to the thought that each law of Nature is based on an underlying symmetry.

Like any other great idea, the symmetry principle should be used with circumspection lest the need of enquiry beyond the search for symmetry is obscured. The hazard lurks therein that nowhere in the world has mathematically precise symmetry ever been encountered. The fundamental symmetries underpinning the laws of Nature, *i.e.* parity (P), charge conjugation (C), and time inversion (T), are hence no more than local approximations and, although the minor exceptions may be just about undetectable, they cannot be ignored[2].

---

[2]Despite known deviations from each of the individual symmetries, it is widely believed by theoreticians that the combined PCT operation is an exact symmetry.

To keep this proposition in focus, equivalence relationships, in the sense of symmetries in the physical world, may be defined in terms of a *metric* for the state space of a system. The metric [14] is a real non-negative function $d(\,,\,)$ with the following properties for all states u, v, w:

$$\text{Self-distance} \quad : \quad d(u, u) = 0$$
$$\text{Symmetry} \quad : \quad d(u, v) = d(v, u)$$
$$\text{Triangle inequality} \quad : \quad d(u, w) \le d(u, v) + d(v, w)$$

A symmetry transformation, or equivalence is introduced by

$$d\big(u, T(u)\big) \le \epsilon$$

The standard formulation of an exact symmetry as

$$d(u, v) = 0 \quad , \quad v = T(u)$$

occurs as a special case only when $\epsilon \to 0$. For non-zero ϵ, however small, all symmetry transformations describe *approximate symmetries* within *approximate symmetry groups*. Note that any approximate symmetry group includes as subgroups all approximate symmetry groups of better approximation, especially the conventional symmetry group ($\epsilon = 0$).

Another term for approximate symmetry is *broken symmetry*. The symmetry breaking factor is whatever factor is responsible for the deviation of ϵ from zero. As an example, any crystal has broken translational symmetry. The exact symmetry limit is an infinite crystal, obviously unattainable in practice.

The whole concept of symmetry and law becomes more palatable against the backdrop of approximate symmetries. Inviolate laws that militate against the scientific spirit, are then prevented by broken symmetries and developments in science amount to relaxing the primitive laws, so as to describe more general situations around the special cases dictated by exact symmetries, *i.e.* by maximizing the parameter ϵ. Any law that reflects a symmetry must then be considered as a useful starting point rather than a final result and conclusions based on perceived symmetries of space and time must be revisited to identify the effects of broken symmetry on the laws of Nature.

The most telling example is the way in which the, often sterile, laws based on the symmetry of special relativity acquire physical significance within the broken symmetry of general relativity. It removes the major anomaly of time-reversible laws of microphysics underpinning reversible macro effects, and shows how local, rather than global gauge-symmetry breaking may cause the creation of massive particles.

To better appreciate the effects of approximate symmetries, the consequences of exact symmetries are examined first.

1.3.1 Conservation Laws

The simplest demonstration of how symmetry fixes natural laws is by the effect of symmetries on the motion of non-relativistic classical particles.

The relativistic, or Lorentz transformation (or *boost*) is a spatiotemporal transformation, which for relative motion along x reads

$$x \to x' = \gamma(x + vt)$$
$$y, z \to y', z' = y, z$$
$$t \to t' = \gamma(t + vx/c^2)$$
where
$$\gamma = \left(1 - v^2/c^2\right)^{-\frac{1}{2}}$$

The non-relativistic analogue, or Galilei transformation is the limiting form of the Lorentz boost as $c \to \infty$ $(v/c \to 0)$, *i.e.*

$$x' = x + vt$$
$$t' = t$$

Conservation of Energy

Assume a potential V that only depends on the coordinate x of the particle under one-dimensional translation. Since V does not depend on time the law that governs evolution of the system must remain constant in time. The total energy

$$E = \frac{1}{2}m\dot{x}^2 + V$$

and the time rate of change

$$\frac{dE}{dt} = m\dot{x}\ddot{x} + \frac{dV}{dx}\dot{x}$$

Substituting the force

$$F = -\frac{dV}{dx} = -m\ddot{x}$$

gives

$$\frac{dE}{dt} = m\dot{x}\ddot{x} - m\ddot{x}\dot{x} = 0$$

Thus the total energy does not change with time and therefore is *conserved*. The conservation of energy clearly is a consequence of the fact that the laws of Nature do not change with time, *i.e.* of their *temporal homogeneity*.

Conservation of Momentum

Conservation of linear momentum is due to spatial homogeneity. Consider i particles of mass m_i at coordinates x_i, as before. Assume an interaction potential that only depends on the separation $x_i - x_j$ between particle pairs. This potential is assumed because it is independent of the location of the system relative to the coordinate origin and the laws governing the evolution of the system are therefore spatial-displacement symmetric. The time rate of change of the total momentum

$$\frac{dP}{dt} = \sum_i m_i \ddot{x}_i$$

where $m_i \ddot{x}_i$ is the force on the ith particle,

$$F_i = -\frac{\partial V}{\partial x_i}$$

Hence

$$\frac{dP}{dt} = -\sum_i \frac{\partial V}{\partial x_i} = -\sum_i \sum_{i \neq j} \frac{\partial V}{\partial (x_i - x_j)}$$

Pairs of terms in the double sum with i and j interchanged cancel in homogeneous space, and therefore

$$\frac{dP}{dt} = 0$$

The implied conservation of momentum is due to the assumed homogeneity of space.

Conservation of Angular Momentum

Assume that the potential of a point particle moving in a plane depends only on its distance from the origin, independent of its orientation with respect to coordinate axes. The angular momentum of the particle with respect to the origin is

$$M = m(x\dot{y} - y\dot{x})$$

$$\frac{dM}{dt} = m(x\ddot{y} - y\ddot{x})$$

Since

$$m\ddot{x} = -\frac{\partial V}{\partial x} = \frac{dV}{dr}\frac{\partial r}{\partial x} = \frac{x}{r}\frac{dV}{dr} \quad , \quad \text{for } r^2 = x^2 + y^2$$

the rate of change of angular momentum

$$\frac{dM}{dt} = \left(-\frac{xy}{r} + \frac{yx}{r}\right)\frac{dV}{dr} = 0$$

assuming rotational symmetry, or isotropy of space.

1.4.1 Symmetry of the Hamiltonian

Since the Laplacian ∇^2 is invariant under orthogonal transformations of the coordinate system [*i.e.* under the 3D rotation-inversion group $O_I(3)$], the symmetry of the Hamiltonian is essentially governed by the symmetry of the potential function V. Thus, if V refers to an electron in a hydrogen atom H would be invariant under the group $O_I(3)$; if it refers to an electron in a crystal, H would be invariant under the symmetry transformations of the space group of the crystal.

Consider the operation P, that corresponds to some coordinate transformation T, on the Schrödinger equation

$$PH\psi = PE\psi$$
$$\text{or} \quad (PHP^{-1})(P\psi) = EP\psi$$
$$\text{i.e.} \quad H'(P\psi) = E(P\psi)$$

where $H' = PHP^{-1}$ is the Hamiltonian referred to the transformed coordinate system. If the operator P is such that $H' = H$, which means that the form of the Hamiltonian function in the new coordinate system is the same as its form in the original system, it is implied that $PH = HP$. This proposition shows that the Hamiltonian commutes with all the operators under which it is invariant. The set of all transformations T which leaves the system invariant forms a group. The set of corresponding operators P leaves the Hamiltonian invariant and hence also constitutes a group. The two groups are isomorphic to each other and is known as the symmetry group of the Hamiltonian. Since $H(P\psi) = E(P\psi)$ it follows that $P\psi$ is also an eigenvector of H with eigenvalue E. The function $P\psi$ is therefore degenerate with ψ, unless it is a multiple thereof. An important conclusion that follows from all this is that for any symmetry (T) of a system there is a corresponding physical observable which remains a constant of motion.

1.4.2 Quantum Invariances

For a Hamiltonian which is invariant under the space inversion operator P it has already been shown that

$$[P, H] = 0 \quad , \quad \frac{dP_H}{dt} = 0 \tag{1.4}$$

Space inversion symmetry therefore yields a conservation law for the physical quantity P, called *parity*. If the state of the system at the given time, is an eigenstate of P belonging to the eigenvalue ± 1, *i.e.* its parity is $+1$ or -1, the system must maintain this parity at any later time.

The effects of parity conservation are not relevant for classical mechanics. Classical systems are in fact arrangements of mixed parity, so that no new information is obtained by taking their mirror images.

The behaviour of a composite quantum system under space inversion may be affected if its constituent particles have *intrinsic* parity. Consider a composite bound system with a Hamiltonian which is invariant under space inversion. Let m_α be the mass of constituent particle α with internal wave function ψ_α, *i.e.*

$$H\psi_\alpha = m_\alpha c^2 \psi_\alpha$$

From (4) follows immediately that

$$HP\psi_\alpha = c^2 m_\alpha P\psi_\alpha$$

and the space inverted eigenstate $P\psi_\alpha$ is seen to have the same energy $m_\alpha c^2$ as ψ_α. If the energy is not degenerate it follows immediately that $P\psi_\alpha$ must be proportinal to ψ_α:

$$P\psi_\alpha = k\psi_\alpha$$

Multiplying the expression by P, remembering that $P^2 = I$, shows that $k = \pm 1$. That is, ψ_α is an eigenvector of P with eigenvalues of ± 1,

$$P\psi_\alpha = \pm\psi_\alpha$$

It is clearly necessary to take account of this property of α to establish the parity of the composite system involving α. The particle α is said to be *scalar* or *pseudoscalar*, with intrinsic parity of $+1$ or -1 if, for space inversion its internal wave function does not, or does change sign.

It should be obvious that the concept of intrinsic parity for a particle is meaningful only if the forces that bind it are invariant under space inversion. The intrinsic parity of a system of particles is defined as the product of the intrinsic parities of the various particles times the parity of the relative orbital angular momenta.

Time reversal in quantum systems also leads to different results compared to classical systems. The basic relationship between the Hamiltonian (or energy) operator and time evolution in quantum mechanics is defined by Schrödinger's equation as

$$H = i\hbar\frac{\partial}{\partial t} \tag{1.5}$$

Although the Hamiltonian does not change sign under time reversal the r.h.s. must obviously do so. Rather than abandon the notion of time reversal as an invariance in quantum theory a different interpretation of the operator I_t may be considered. Under the usual formulation

$$\psi(\boldsymbol{x}, t) \xrightarrow{I_t} \psi'(\boldsymbol{x}, t') = \eta\psi(\boldsymbol{x}, t)$$

the Schrödinger equation for a free particle

$$i\hbar\frac{\partial}{\partial t}\psi(\boldsymbol{x},t) = -\frac{\hbar^2}{2m}\nabla^2\psi(\boldsymbol{x},t) \tag{1.6}$$

becomes

$$i\hbar\frac{\partial}{\partial t'}\psi'(\boldsymbol{x},t') = -i\hbar\frac{\partial}{\partial t}\eta\psi(\boldsymbol{x},-t) = \frac{\hbar^2}{2m}\nabla^2\eta\psi(\boldsymbol{x},-t) = \frac{\hbar^2}{2m}\nabla^2\psi'(\boldsymbol{x},t')$$

and the invariance is lost. The remedy is to compensate for the sign change by taking the complex conjugation of (6) when replacing t by $-t$, *i.e.*

$$i\hbar\frac{\partial\psi^*(\boldsymbol{x},-t)}{\partial t} = -\frac{\hbar^2}{2m}\nabla^2\psi(\boldsymbol{x},-t)$$

This equation has the same form as (6). Hence if $\psi(\boldsymbol{x},t)$ is a solution of (6), $\psi^*(\boldsymbol{x},-t)$ is also a solution, known as the *time-reversal* solution. In other words, the Schrödinger equation will be invariant under time reversal represented by the mapping

$$\psi(\boldsymbol{x},t) \underset{\longrightarrow}{I_t} \psi'(\boldsymbol{x},t') = \eta\psi^*(\boldsymbol{x},-t)$$

where η is an arbitrary phase factor, $|\eta|^2 = 1$. It turns out that such an *anti-linear* transformation is an acceptable operation since the absolute value of scalar products that represent measurable quantities is preserved, $\langle\phi|\psi\rangle^* = \langle\psi|\phi\rangle$. Schrödinger's equation (5) is therefore seen to stay invariant under temporal inversion since I_t as defined above, causes the sign of both i and dt to change on the r.h.s.

1.5 Field Symmetries

Special techniques are required to describe the symmetry of fields. Since fields are defined in terms of continuous variables it is desirable to formulate suitable transformations of dynamic variables pertaining to fields, in terms of continuous parameters. This is done by using Hamilton's principle and defining quantities such as momentum densities for any field. The most useful parameter to quantify the symmetry of a field is the Lagrangian density (T 3.3.1).

In the discrete case the Lagrangian is a function of the coordinates q_i and the generalized velocities $\dot{q}_i = dq_i/dt$.

A typical field may be defined as $\phi(\boldsymbol{x},t)$ where there is one generalized coordinate for each point $\boldsymbol{x}$ of space. Familiar examples are the electric and magnetic fields, $\boldsymbol{E}$ and $\boldsymbol{B}$. The Lagrangian must evidently involve an integral over the continuum labels $\boldsymbol{x}$, thus

$$L = \int d^3x \mathcal{L}$$

where $\mathcal{L}$ is called the Lagrangian density, which is a function of ϕ, $d\phi/dt$, and the coordinate differences $\nabla\phi$. Using Minkowski notation where

$$\partial_\mu \phi \equiv \frac{\partial \phi}{\partial x^\mu} \quad (\mu = 0, 1, 2, 3),$$

$$\mathcal{L} = \mathcal{L}(\phi, \partial_\mu \phi) \quad \text{with } x^0 = ct.$$

The action $S[\phi]$ is a functional of the field, $\phi(x)$,

$$S[\phi] = \int_{t_1}^{t_2} L dt = \int_{t_1}^{t_2} d^4x \mathcal{L}(\phi, \partial_\mu \phi).$$

Lagrange's equations follow by demanding that S is stationary under variations of the generalized coordinates q. Thus, for an arbitrary change in the field ϕ:

$$\phi \to \phi + \delta\phi$$

the action changes by

$$S \to S + \delta S$$

where

$$
\begin{aligned}
\delta S &= \int_{t_1}^{t_2} d^4x \left(\frac{\partial \mathcal{L}}{\partial \phi} \delta\phi + \frac{\partial \mathcal{L}}{\partial(\partial_\mu \phi)} \delta(\partial_\mu \phi) \right) \\
&= \int_{t_1}^{t_2} d^4x \left[\frac{\partial \mathcal{L}}{\partial \phi} - \partial_\mu \left(\frac{\partial \mathcal{L}}{\partial(\partial_\mu \phi)} \right) \right] \delta\phi + \int_{t_1}^{t_2} d^4x \partial_\mu \left(\frac{\partial \mathcal{L}}{\partial(\partial_\mu \phi)} \delta\phi \right) \quad (1.7)
\end{aligned}
$$

Integration of the last term of (7) involves the value of the quantity in brackets on the surface of the hypersphere only. If the variation of $\delta\phi$ is also restricted to this surface the last term vanishes and the functional derivative

$$\frac{\delta S}{\delta \phi(x)} = \frac{\partial R}{\partial \phi} - \partial_\mu \left(\frac{\partial \mathcal{L}}{\partial(\partial_\mu \phi)} \right)$$

which vanishes for stationary S to give the Euler-Lagrange equations

$$\frac{\partial \mathcal{L}}{\partial \phi} - \partial_\mu \left(\frac{\partial \mathcal{L}}{\partial(\partial_\mu \phi)} \right) = 0 \qquad (1.8)$$

It is common practice to refer to the Lagrangian density simply as the Lagrangian of the field under the same symbol L.

Lagrangian densities can be constructed to yield any desired field equation. For example

$$L = \frac{1}{2}\rho(\partial^\mu y)(\partial_\mu y)$$

with ρ a mass density, gives

$$\frac{\partial L}{\partial y} = 0 \quad \text{and} \quad \frac{\partial L}{\partial(\partial_\mu y)} = \rho\partial^\mu y$$

so (8) implies

$$\Box y \equiv \partial_\mu \partial^\mu y = \frac{1}{c^2}\frac{\partial^2 y}{\partial t^2} - \nabla^2 y = 0$$

which is the wave equation.

1.5.1 Noether's Theorem

An advantage of the Lagrangian formulation is that it permits the identification of conserved quantities by studying the invariances of the action S. The fundamental result behind this statement is Noether's theorem, which identifies conserved currents associated with invariance of S (and hence L) under infinitesimal continuous transformation.

Consider a general transformation of coordinates x and field ϕ. Suppose

$$x^\mu \rightarrow (x')^\mu = x^\mu + \delta x^\mu$$

$$\text{and} \quad \phi(x) \rightarrow \phi'(x') = \phi(x) + \delta\phi(x)$$

Both of these transformations can be formulated in terms of a common set of infinitesimal parameters $\delta\omega^i$, *i.e.*

$$\delta x^\mu = X_i^\mu(x)\delta\omega^i \tag{1.9}$$
$$\delta\phi(x) = \Phi_i(x)\delta\omega^i \tag{1.10}$$

The variation $\delta\phi(x)$ contains contributions of variations in both the field function and its argument

$$\begin{aligned}
\phi'(x') &= \phi'(x + \delta x) \\
&= \phi'(x) + \delta x^\nu(\partial_\nu\phi) \\
&= \phi(x) + \delta_0\phi(x) + \delta x^\nu(\partial_\nu\phi)
\end{aligned} \tag{1.11}$$

where $\delta_0\phi$ is the variation of the field function alone. Equations (9), (10) and (11) can be combined into

$$
\begin{aligned}
\delta_0\phi(x) &= \delta\phi(x) - \delta x^\nu(\partial_\nu\phi) \\
&= [\Phi_i(x) - (\delta_\nu\phi)X_i^\nu(x)]\delta\omega_i
\end{aligned}
$$

The variation of S

$$
\delta S = \int_{t_1}^{t_2} \delta(d^4x)L + \int_{t_1}^{t_2} \delta L(d^4x) \tag{1.12}
$$

where $\delta(d^4x)$ is the variation of the integration measure caused by (9). It can be shown [15] that this variation can be written as

$$
\delta(d^4x) = [\partial_\mu(X_i^\mu\delta\omega^i)]d^4x
$$

If ϕ satisfies the Euler-Lagrange equation (8) only the second integral on the r.h.s. of (7) survives, and hence

$$
\int_{t_1}^{t_2} \delta L d^4x = \int_{t_1}^{t_2} d^4x \left[\frac{\partial L}{\partial x^\mu} + \partial_\mu\left(\frac{\partial L}{\partial(\partial_\mu\phi)}\delta_0\phi\right)\right]
$$

Expressing δx^μ and $\delta_0\phi$ in terms of $\delta\omega^i$ and substituting the previous two results into (12) give

$$
\delta S = \int_{t_1}^{t_2} d^4x\partial_\mu\left(LX_i^\mu\delta\omega^i + \frac{\partial L}{\partial(\partial_\mu\phi)}[\Phi_i - (\partial_\nu\phi)X_i^\nu]\delta\omega^i\right)
$$

If S is invariant with variations parametrized by *constant* $\delta\omega^i$, S is said to be invariant under *global* transformation,

$$
\frac{\delta S}{\delta\omega^i} = 0
$$

and the quantity

$$
j_i^\mu \equiv \left(\frac{\partial L}{\partial(\partial_\mu\phi)}(\partial_\nu\phi) - L\delta_\nu^\mu\right)X_i^\nu - \frac{\partial L}{\partial(\partial_\mu\phi)}\Phi_i \tag{1.13}
$$

is conserved (Note that δ_ν^μ is a delta function and not a variation):

$$
\partial_\mu j_i^\mu = 0.
$$

This is Noether's theorem. The quantity j_i^μ in the parlance of modern physics is known as a conserved current. The theorem (13) states that if the action of a system (or the Lagrangian density) is invariant under transformations of type (9) and (10) then there exist i conserved quantities.

The conservation of energy, momentum and angular momentum, considered before, in all cases is a consequence of Noether's theorem [17]. An important new result is obtained by applying the theorem to internal symmetry transformations. This result is illustrated by the theory of the complex field[4]

$$\phi(x) = \frac{1}{\sqrt{2}}[\phi_1 + i\phi_2(x)]$$

where $\phi(x)(i = 1, 2)$ are real scalar fields. The Lagrangian density is a function of both fields and their derivatives, or equivalently of $\phi(x)$, its complex conjugate $\phi^*(x)$, and derivatives:

$$L = (\partial_\mu \phi^*)(\partial^\mu \phi) - \mu^2 \phi^* \phi - \lambda(\phi^* \phi)^2$$

In the absence of interaction $\lambda = 0$, and the Euler-Lagrange condition implies the Klein-Gordon equation

$$(\Box + \mu^2)\phi(x) = 0$$

The Lagrangian is invariant under the global gauge transformation

$$\phi(x) \rightarrow \phi'(x) = e^{-iq\Lambda}\phi(x)$$

or

$$\phi^*(x) \rightarrow (\phi^*)'(x) = e^{iq\Lambda}\phi^*(x)$$

For infinitesimal Λ

$$e^{-iq\Lambda} \approx 1 - iq\Lambda$$

and $\delta\phi(x) = -iq\Lambda\phi(x)$ From(10) $\Phi(x) = -iq\phi(x)$. Similarly $\delta(\phi^*) = (\delta\phi)^*$, and so $\Phi^*(x) = [\Phi(x)]^*$.

Since S, like L, is invariant the Noether current

$$
\begin{aligned}
j^\mu &= -\frac{\partial L}{\partial(\partial_\mu \phi)}\Phi - \frac{\partial L}{\partial(\partial^\mu \phi^*)}\Phi^* \\
&= (\partial^\mu \phi^*)iq\phi - (\partial^\mu \phi)iq\phi^*
\end{aligned}
$$

is conserved, $\partial_\mu j^\mu = 0$. It follows that the charge

$$Q = \int d^3x\, j^0$$

---

[4]The electromagnetic field is an example of a complex vector field, $\boldsymbol{Q} = \boldsymbol{B} + (i/c)\boldsymbol{E}$.

is conserved because of the $U(1)$ phase invariance,

$$\frac{dQ}{dt} = \int d^3x \partial_0 j^0(\boldsymbol{x}, t) = 0$$

1.5.2   Symmetry Breaking

The Lagrangian formulation is equally effective for discussing the idea of spontaneous breakdown of symmetry and the effect thereof can be demonstrated in terms of a simple example [18, 19].

Consider a system consisting of scalar particles only and described by a Lagrangian density as a function of the field ϕ. Let

$$L = T - V = \frac{1}{2}(\partial_\mu \phi)^2 - \left(\frac{1}{2}\mu^2\phi^2 + \frac{1}{4}\lambda\phi^4\right) \tag{1.14}$$

with $\lambda > 0$. This Lagrangian is invariant under the symmetry operation that replaces ϕ by $-\phi$. The two possible forms of the potential field correspond closely to the two curves in figure 3. The upper curve describes a scalar

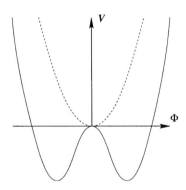

Figure 1.3: *The potential* $V(\phi) = \frac{1}{2}\mu^2\phi^2 + \frac{1}{4}\lambda\phi^4$ *for* $\mu^2 > 0$ *(upper curve) and* $\mu^2 < 0$, $\lambda > 0$.

field with mass μ. The ϕ^4 term describes self-interaction of the field. The ground state corresponds to $\phi = 0$. It obeys the reflection symmetry of the Lagrangian.

The lower curve corresponds to $\mu^2 < 0$ and appears to have negative mass for the field ϕ. The potential has two minima which satisfy

$$\frac{\partial V}{\partial \phi} = \phi\left(\mu^2 + \lambda\phi^2\right)$$

and are therefore at

$$\phi = \pm v \quad \text{with} \quad v = \sqrt{-\mu^2/\lambda}.$$

The extremum $\phi = 0$ does not correspond to the energy minimum, which occurs at $\phi = \pm v$. If $\eta(x)$ represents fluctuations around the minimum at v, one can write

$$\phi(x) = v + \eta(x)$$

If this form of the field, translated to $\phi + v$, is now substituted into the Lagrangian it becomes

$$L' = \frac{1}{2}(\partial_\mu \eta)^2 - \lambda v^2 \eta^2 - \lambda v \eta^3 - \frac{1}{4}\lambda \eta^4 + \frac{1}{4}v^4 \lambda$$

and appears to be different from (14), although the two forms should be equivalent in all respects. The most important difference is that the mass term, $-\lambda v^2 \eta^2$, of the η field now has the correct sign; $m_\eta = \sqrt{2\lambda v^2} = \sqrt{-2\mu^2}$. The difference between the two Lagrangians is that L' has been evaluated at the true minimum whereas L was expanded around an unstable point at $\phi = 0$. L' represents the situation correctly and is said to have *generated*, or better, *revealed* the mass by spontaneous symmetry breaking. The reflection symmetry of the Lagrangian has apparently been broken by choosing the ground state $\phi = v$, rather than $\phi = -v$.

One example of a system in which the ground state does not possess the symmetry of the Lagrangian is the buckling of a compressed knitting needle shown in figure 5. When vertical pressure on the needle is increased

Figure 1.4: *Symmetry breaking by pressure applied to a cylindrically symmetrical needle.*

to some critical level, the needle buckles and snaps into an arrangement of lower symmetry by randomly selecting a plane into which it curves. The cylindrical symmetry of the system is apparently broken by the buckling of the needle.

The Goldstone theorem

Another important feature of symmetry breaking is revealed when applied to continuous symmetries. An example is the complex scalar field

$$\phi = \frac{1}{\sqrt{2}}(\phi_1 + i\phi_2)$$

which is described by the Lagrangian

$$L = (\partial_\mu \phi)^*(\partial^\mu \phi) - \mu^2 \phi^* \phi - \lambda(\phi^* \phi)^2 \qquad (1.15)$$

and which is invariant under the global transformation

$$\phi \to e^{i\alpha}\phi$$

of the symmetry group $U(1)$.

As before, consider the case when $\lambda > 0$ and $\mu^2 < 0$. When rewriting the Lagrangian in the form

$$L = \frac{1}{2}(\partial_\mu \phi_1)^2 + \frac{1}{2}(\partial_\mu \phi_2)^2 - \frac{1}{2}\mu^2(\phi_1^2 + \phi_2^2) - \frac{1}{4}\lambda(\phi_1^2 + \phi_2^2)^2$$

there is seen to be a circle of minima of the potential $V(\phi)$ in the ϕ_1, ϕ_2 plane and of radius v, such that

$$\phi_1^2 + \phi_2^2 = v^2 \quad , \quad v^2 = -\frac{\mu^2}{\lambda} \quad ,$$

as shown in figure 4. As before, the field ϕ is translated to a minimum energy

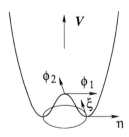

Figure 1.5: *The potential $V(\phi)$ of a complex scalar field for the case $\mu^2 < 0$ and $\lambda > 0$.*

position, which without loss of generality may be taken as the point $\phi_1 = v$,

$\phi_2 = 0$. The Lagrangian is expanded at this point in terms of the fields η and ξ by substituting

$$\phi(x) = \frac{1}{\sqrt{2}}[v + \eta(x) + i\xi(x)]$$

into (15) to give

$$L' = \frac{1}{2}(\partial_\mu \xi)^2 + \frac{1}{2}(\partial_\mu \eta)^2 + \mu^2 \lambda^2 + \text{const.} + \text{cubic and quartic terms in } \eta, \xi.$$

The third term has the form of a mass term $(-\frac{1}{2}m_\mu^2 \eta^2)$ for the η field. Thus, the η-mass is $m_\eta = \sqrt{-2\mu^2}$, as before. The first term in L' represents the kinetic energy of the ξ-field, but there is no mass term for ξ. This means that the ξ field is carried by massless particles, known as Goldstone bosons.

The appearance of the Goldstone boson is not too difficult to understand. The potential in the direction of ξ is flat and there is no resistance to excitations, which implies a massless mode. The Lagrangian of the complex field is one example of the Goldstone theorem, which states that massless scalars occur whenever a continuous symmetry of a physical system is spontaneously broken. A more accurate statement would be that this happens when the global symmetry is not apparent in the ground state. The massless boson mediates the interaction that breaks the symmetry. A well-known example of this phenomenon is a ferromagnet, where the spin alignment in the ground state corresponds to a special direction which is selected at random when the transition occurs from a state of higher symmetry and randomly aligned spins. The Goldstone boson in this case is the long-range spin waves which are oscillations of the spin alignment.

1.6 Symmetry in Theories

Not only the laws of Nature but also all major scientific theories are statements of observed symmetries. The theories of special and general relativity, commonly presented as deep philosophical constructs can, for instance, be formulated as representations of assumed symmetries of space-time. Special relativity is the recognition that three-dimensional invariances are inadequate to describe the electromagnetic field, that only becomes consistent with the laws of mechanics in terms of four-dimensional space-time. The minimum requirement is euclidean space-time as represented by the symmetry group known as Lorentz transformation.

A more general formulation in terms of a non-euclidean manifold (curved space) has several advantages, the most important of which is a geometrical model of gravitation. A common assumption is that, locally the general theory reduces to the euclidean Lorentz group, which consequently, by definition, describes an approximate symmetry. In order to define the electromagnetic and gravitational fields within the same theory it is necessary to consider the five-dimensional world symmetry of Kaluza and Klein.

In other theories the symmetry connection may be less obvious. It will be noticed that most of the symmetries pertaining to laws and theories are described by continuous groups. The reason is that all of these theories and laws describe processes rather than structure. Theories pertaining to structures are expected to arise from symmetries better formulated in terms of discrete groups. Theories of molecular structure, just like the structures themselves, should be described by point-group symmetries and theories of the solid state should be based on crystallographic point groups and Laue symmetry. At the interface between processes and structures are transformations, such as structural phase transitions. These are best described in terms of symmetry groups with both continuous and discrete components, such as Bloch functions and space groups. Especially the four-dimensional space groups can be anticipated to gain importance as a vehicle for future theories of structural transformations. Selected symmetry groups of theoretical importance are discussed in the following.

1.6.1   Continuous Translational Symmetry

Consider a particle located at position x_0 on the real line and the translation operator T that causes displacement of the particle by a distance x. To avoid confusion the position x_0 may be defined as the particle state $|x_0\rangle$, using Dirac notation, although the particle may be classical. The action of $T(x)$ on x_0 is written

$$T(x)|x_0\rangle \equiv |x + x_0\rangle$$

It is easy to see that $T(x)$ must have the following properties:

$$\begin{aligned}
T(x_1)T(x_2) &= T(x_1 + x_2) \\
T(0) &= E \\
T^{-1}(x) &= T(-x)
\end{aligned}$$

These properties are just those required for $T(x)$ in the interval $-\infty < x < +\infty$ to constitute a group. Moreover, since the translation is a continuous operation the homorphism

$$f(x + y) = f(x)f(y)$$

is *analytic* and each element an image in a topological space. Such a set, which is both manifold and group, is known as a Lie group, *i.e.* an infinite group whose parameters can be parametrized smoothly and analytically. This property implies that an element of the group in the neighbourhood of the identity, conveniently located at the origin of the coordinate space, is generated by an infinitesimal operator, $(df)_0(1)$, say. By extending x to the $\mathbb{R}$ manifold

$$f(x) \to f(x/N)^N$$

for suitably large N. This function is a one-parameter subgroup f_v of v. The map $v \to f_v$ is the inverse of the map $(df)_0(1)$. Using the algebraic identity

$$f\left(1 + \frac{x}{N}\right)^N = \exp(x)$$

and taking $d(\exp)_0$ as the identity map, any one-parameter subgroup can be formulated [16] as

$$f_v(x) = \exp(xv)$$

in terms of the *generator* v. A representation of the translational Lie group with unitary operators becomes

$$T(x) = e^{-iPx}$$

where the generator P corresponds to a Hermitian operator with real eigenvalues, p say. For the representation $T(x) \to U^p(x)$ follows

$$P|p\rangle = p|p\rangle$$

$$U^p(x)|p\rangle = |p\rangle e^{-ipx}$$

The localized states $|x\rangle$ and the translational states $|p\rangle$ are related by an expression such as

$$|p\rangle = \int_{-\infty}^{\infty} |x\rangle e^{-ipx} dx$$

which is the well-known quantum-mechanical expression that relates momentum states and coordinate states. The translational operator may therefore be written as

$$T(x) = \exp(-ipx/\hbar)$$

$$p = -i\hbar\nabla$$

the quantum mechanical momentum operator.

1.6.2   Continuous Rotational Symmetry

A system symmetric under rotations around a fixed point in a plane is described by the rotation group $O(2)$. An arbitrary vector $\boldsymbol{x}$ in the plane transforms under rotation $R(\phi)$ according to

$$\boldsymbol{x} \rightarrow \boldsymbol{x}' \equiv \boldsymbol{R}(\phi)\boldsymbol{x}$$

which is satisfied by

$$R(\phi) = \begin{pmatrix} \cos\phi & -\sin\phi \\ \sin\phi & \cos\phi \end{pmatrix}$$

The length of the vector $|x|^2$ remains invariant under rotation and it is easy to show that $R(\phi)R^T(\phi) = E$ $\forall \phi$, where R^T is the transpose of R and E is the unit matrix. Real matrices that satisfy this condition are known as *orthogonal* matrices. The condition implies that $[\det R(\phi)]^2 = 1$ or that $\det R(\phi) = \pm 1$. Matrices with determinant equal to -1 correspond to rotations combined with spatial inversion or mirror reflection. For pure rotations $\det R = 1$, for all ϕ.

Two rotations in succession result in an equivalent single rotation, with the obvious law of composition

$$R(\phi_2)R(\phi_1) = R(\phi_2 + \phi_1)$$

with the understanding that whenever $\phi_1 + \phi_2$ goes outside the range $(0, 2\pi)$,

$$R(\phi) = R(\phi \pm 2\pi)$$

The group elements of $O(2)$ are all $0 \leq \phi \leq 2\pi$, the points on the unit circle that defines the topology of the group parameter space.

An infinitesimal rotation by the angle $d\phi$ for analytic $R(\phi)$ is defined, as before by

$$R(d\phi) \equiv E - iJd\phi$$

such that

$$\frac{dR(\phi)}{d\phi} = -iJR(\phi)$$

with the boundary condition $R(0) = E$. The unique solution defines the generator J of the group through

$$R(\phi) = e^{-iJ\phi}$$

Since $O(2)$ is an abelian group, all its irreducible representations are one-dimensional. For any vector $|\alpha\rangle$

$$J|\alpha\rangle = |\alpha\rangle\alpha$$

$$U(\phi)|\alpha\rangle = |\alpha\rangle e^{-i\phi\alpha}$$

where α is the eigenvalue of the Hermitian operator J. The condition that $\exp(\pm 2\pi i\alpha) = 1$ implies that α is an integer, usually denoted by m. The rotational states

$$|\phi\rangle = \sum_m |m\rangle e^{im\phi}$$

transformed under the operator J are

$$J|\phi\rangle = \sum_m J|m\rangle e^{-im\phi} = \sum_m |m\rangle m e^{-im\phi} = i\frac{d}{d\phi}|\phi\rangle,$$

and J may be recognized as the quantum mechanical angular momentum operator in units of $\hbar$.

1.6.3   Discrete Rotational Symmetry

The operators of discrete rotational groups, best described in terms of both proper and improper symmetry axes, have the special property that they leave one point in space unmoved; hence the term *point group*. Proper rotations, like translation, do not affect the internal symmetry of an asymmetric motif on which they operate and are referred to as operators of the first kind. The three-dimensional operators of improper rotation are often subdivided into *inversion axes, mirror planes* and *centres of symmetry*. These operators of the *second kind* have the distinctive property of inverting the handedness of an asymmetric unit. This means that the equivalent units of the resulting composite object, called *left* and *right*, cannot be brought into coincidence by symmetry operations of the first kind. This inherent handedness is called *chirality*.

1.6.4   Lattice Translations

The translational symmetry of a (one-dimensional) crystal lattice demands that the potential in the crystal Hamiltonian

$$H = \frac{p^2}{2m} + V(x)$$

should satisfy the periodicity condition

$$V(x + na) = V(x), \quad \text{for all } n = \text{ an integer.}$$

The Hamiltonian must be invariant under translations along the lattice by any integral multiple (n) of the lattice spacing a,

$$x \to x' = x + na.$$

For a quantum particle the system is described by a state vector $|\psi\rangle$ which transforms into $|\psi'\rangle$ after symmetry translation, as described by the operator $T(n)$, in the vector space of physical states Φ

$$|\psi\rangle \to |\psi'\rangle = T_n|\psi\rangle, \quad \forall|\psi\rangle \in \Phi$$

Since this is a symmetry operation, the two sets of vectors $\{|\psi\rangle\}$ and $\{|\psi'\rangle\}$ for any given $T(n)$, must provide equivalent descriptions of the physical system, and $T(n)$ must be a linear operator. In addition, all physical observables must remain invariant under the symmetry operation. Physical observables are expressed in terms of scalar products, such as $\langle\phi|\psi\rangle$ and the probability that a system described by $|\psi\rangle$ will be found in state $|\phi\rangle$ must be unchanged by $T(n)$, i.e.

$$|\langle\phi|\psi\rangle|^2 = |\langle\phi'|\psi'\rangle|^2 = |\langle\phi|T(n)^\dagger T(n)|\psi\rangle|^2,$$

and so $T(n)$ must be a unitary operator. The set of symmetry operators on the lattice is said to be *realized* on the vector space Φ by the set of unitary operators $\{T(n)\}$. Alternatively, the operators $\{T(n)\}$ are said to form a *representation* of the symmetry operators of the Hamiltonian.

Bloch Functions

The translational operators are required by physical principles and simple geometry to satisfy the following conditions:

$$\begin{aligned} T(n)T(m) &= T(n+m) \\ T(0) &= E \\ T(-n) &= T(n)^{-1} \end{aligned}$$

These properties define a group. This group of discrete translations will be denoted by T^d. Two additional properties are worth noting

$$T(n)T(m) = T(m)T(n)$$

$$T^{\dagger}(n)T(n) = E$$

Since all $T(n)$ commute with each other a set of basis vectors in Φ can be chosen to be simultaneous eigenvectors of $T(n)$ for all n. The members of the basis are denoted by $|u(\xi)\rangle$ where ξ is a yet unspecified label for the vectors. This means that

$$T(n)|u(\xi)\rangle = |u(\xi)\rangle t_n(\xi)$$

where $t_n(\xi)$ are the eigenvalues of $T(n)$ corresponding to the eigenvector $|u(\xi)\rangle$. Applying the five basic conditions and invoking the linear independence of $|u(\xi)\rangle$ for distinct ξ, give the following five relationships in terms of eigenvalues:

$$
\begin{aligned}
t_n(\xi)t_m(\xi) &= t_{n+m}(\xi) \\
t_0(\xi) &= 1 \\
T_{-n}(\xi) &= \frac{1}{t_n(\xi)} \\
t_n(\xi)t_m(\xi) &= t_m(\xi)t_n(\xi) \\
|t_n(\xi)|^2 &= 1
\end{aligned}
$$

The last condition implies

$$t_n(\xi) = e^{-i\phi_n(\xi)}$$

where $\phi_n(\xi)$ are real numbers.

The first three of the relations then become

$$
\begin{aligned}
\phi_n(\xi) + \phi_m(\xi) &= \phi_{n+m}(\xi) \\
\phi_0(\xi) &= 0 \\
\phi_{-n}(\xi) &= -\phi_n(\xi)
\end{aligned}
$$

The solution of these three equations is

$$\phi_n(\xi) = nf(\xi),$$

where f is an arbitrary function. Since ξ still has to be specified it may be chosen such that $f(\xi) = \xi$ and $\phi_n(\xi) = n\xi$, which leads to

$$t_n(\xi) = e^{-in\xi} \tag{1.16}$$

The eigenvalues form a representation of the discrete translational group and equation (16) indicates that $t_n(\xi)$ is periodic in the variable ξ with period

2π, e.g. $-\pi \leq \xi \leq \pi$. In terms of the new parameter $\boldsymbol{k}= \xi/a$ (the wave vector)

$$T_n|u(E, \boldsymbol{k})\rangle = |u(E, \boldsymbol{k})\rangle e^{-i\boldsymbol{k}na}$$

for eigenvalues E of the Hamiltonian equation

$$H|u(E, \boldsymbol{k})\rangle = |u(E, \boldsymbol{k})\rangle E$$

The wave function in coordinate space (*i.e* the Schrödinger wave function)

$$u_{E,\boldsymbol{k}}(x) = \langle x|u(E, \boldsymbol{k})\rangle$$

However, since x must be periodic it may be specified as

$$(x = y + na) \quad -\frac{a}{2} \leq y \leq \frac{a}{2}$$

Like the pointer on the clock it thus always ends up in the same one-dimensional box, or *unit cell*. Since

$$T(n)|y\rangle = |y + na\rangle$$

it follows directly that $T(n)|y\rangle = |x\rangle$ and hence $\langle x| = \langle y|T^\dagger(n)$. The wave function can therefore be written as

$$
\begin{aligned}
u_{\boldsymbol{k}}(x) &= \langle y|u(\boldsymbol{k})\rangle e^{i\boldsymbol{k}na} \\
&= \langle y|T(-n)|u(\boldsymbol{k})\rangle \\
&= \langle y|u(\boldsymbol{k})\rangle e^{i\boldsymbol{k}na} \\
&= u_{\boldsymbol{k}}(y)e^{i\boldsymbol{k}(x-y)}
\end{aligned}
$$

Therefore

$$u_{\boldsymbol{k}}(x)e^{-i\boldsymbol{k}x} = u_{\boldsymbol{k}}(y)e^{-i\boldsymbol{k}y}$$

for all y and $-\frac{a}{2} \leq x \leq \frac{a}{2}$.

A new wave function, periodic in x, with periodicity a is therefore defined by

$$\phi(x) = u_{\boldsymbol{k}}(x)e^{-i\boldsymbol{k}x}$$

This statement is known as Bloch's theorem, which in three dimensions reads

$$\phi(\boldsymbol{r}) = u_{\boldsymbol{k}}(\boldsymbol{r}) \exp(i\boldsymbol{k}.\boldsymbol{r})$$

or equivalently

$$\phi(\boldsymbol{r} + \boldsymbol{t}) = \phi(\boldsymbol{r}) \exp(i\boldsymbol{k}.\boldsymbol{r})$$

for a translation t.

The boundary conditions are periodic and the number of allowed values of the wave vector $\boldsymbol{k}$ is equal to the number of unit cells in the crystal. These eigenfunctions constitute the basis for the infinite-dimensional Hilbert space of the crystal Hamiltonian and any function with the same boundary conditions can be expressed as a linear combination of functions in this complete set.

1.6.5 Crystallographic Symmetry

The clock example illustrates most principles of importance in discrete symmetry groups with translation, also known as crystallographic symmetry groups. For simplicity consider a two-dimensional unit cell with a two-fold axis T, pictured as a pointed ellipse, and two mirror planes m_V and m_H. To construct a multiplication table any general position (not coincident with

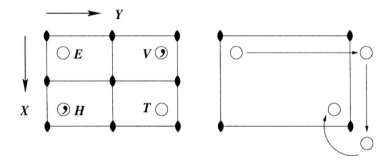

Figure 1.6: *Two-dimensional space group p2mm*

one of the symmetry elements) is chosen to represent the identity operator, E, at coordinates x, y. Each of the other symmetry elements T, V and H are allowed to operate on this general point and move it to the sites T, V, H respectively. The multiplication table is:

G	E	T	V	H
E	E	T	V	H
T	T	E	H	V
V	V	H	E	T
H	H	V	T	E

It is easy to show that the table remains the same while labelling the sites differently, starting from different sites for E. It is therefore more practical to label the sites according to the coordinates generated by operating on the x, y site, $i.e.$

$$E \rightarrow x, y$$

$$T \rightarrow \bar{x}, \bar{y}$$

$$V \rightarrow x, \bar{y}$$

$$H \rightarrow \bar{x}, y$$

Several important concepts emerge from this simple model:

1. Translational symmetry is assumed in both directions, X and Y. This assumption implies that if an operator moves any site into a neighbouring cell it is equivalent to that site entering the reference cell from the opposite side. The coordinates x, y are therefore fractional and symmetry translations are unity, whereby $1 + x \rightarrow x$; $\bar{x} = 1 - x$, $etc.$

2. Each symmetry element interacts with all others to generate operators at equivalent sites commensurate with group composition. It is readily demonstrated diagramatically how the elements $2\,m\,m$ interact with translational elements, $e.g.$ using two translations and a rotation to generate the equivalent site $\bar{x}, \bar{y}$ from x, y. However, the equivalence of these two sites generates a further two-fold axis at $1/2, 1/2$. Comparison of all equivalent sites generates a complete set of elements which ensures that the group operations are all possible within one unit cell.

3. Operations of the second kind are identified by commas, noting that a given operator that transforms an empty circle into a circled comma also does the opposite.

4. The number of equivalent general positions (the *multiplicity*) is the same as the number of symmetry elements.

5. For certain choices of starting positions two or more symmetry-equivalent sites may merge into one, like the sites on one of the mirror planes, $e.g.$ $1/2, y$.

The equivalent positions become

$$1/2, y \,;\, 1/2, \bar{y} \,;\, 1/2, y \,;\, 1/2, \bar{y}$$

which reduces to the two sites $1/2, y$ and $1/2, \bar{y}$. This type of site is called a *special position*, with half multiplicity of 2. The sites $0, 0$; $1/2, 1/2$; $etc.$ have

Chapter 2

Mathematical Structures in Chemistry

2.1 Introduction

The broken symmetries of chapter 1, assumed responsible for shaping the physical world, refer to the symmetry of the vacuum and thus finally to the geometry of space-time. It is not immediately obvious that chemical theories could also be reduced to the same cause. While physics produced quantum theory and general relativity, the fundamental contribution from chemistry was the periodic table of the elements. Although the structure of individual atomic nuclei may be considered shaped by local space-time symmetry, the functional relationship between different nuclides needs further study.

The modern view of the periodic table explains its structure in terms of an Aufbau procedure based on the wave-mechanical model of the hydrogen atom. Although seductive at first glance, the model is totally inadequate to account for details of the observed electronic configurations of atoms, and makes no distinction between isotopes of the same element. The attractive part of the wave-mechanical model is that it predicts a periodic sequence of electronic configurations readily specified as a function of atomic number. The periodicity follows from the progressive increase of four quantum numbers n, l, m_l and s, such that

$$n = 1, 2, 3 \ldots$$
$$l = 0, 1 \ldots, (n-1)$$
$$m_l = -l, \ldots, +l$$
$$s = \pm \frac{1}{2}$$

In terms of the exclusion principle each electron of a given atom has a unique combination of the four quantum numbers. The Aufbau principle assumes that the quantum numbers increase by unit increments in the order: $s \rightarrow m_l \rightarrow l \rightarrow n$, starting from the minimum set $(-\frac{1}{2}, 0, 0, 1)$, with increasing atomic number, Z. This procedure assumes that $(Z-1)$ electrons on each atom have the same quantum numbers as those on the previous atom in the sequence. The additional electron that raises the atomic number to Z has the quantum number shown for the first few elements in the periodic table, in table 1. This prescription reproduces the correct sequence and periodic

Z	Atom	n	l	m_l	s	Symbol
1	H	1	0	0	$-\frac{1}{2}$	$1s^1$
2	He	1	0	0	$+\frac{1}{2}$	$1s^2$
3	Li	2	0	0	$-\frac{1}{2}$	$1s^2 2s^1$
4	Be	2	0	0	$\frac{1}{2}$	$1s^2 2s^2$
5	B	2	1	-1	$-\frac{1}{2}$	$1s^2 2s^2 2p^1$
6	C	2	1	-1	$\frac{1}{2}$	$1s^2 2s^2 2p^2$
7	N	2	1	0	$-\frac{1}{2}$	$1s^2 2s^2 2p^3$
8	O	2	1	0	$\frac{1}{2}$	$1s^2 2s^2 2p^4$
9	F	2	1	1	$-\frac{1}{2}$	$1s^2 2s^2 2p^5$
10	Ne	2	1	1	$\frac{1}{2}$	$1s^2 2s^2 2p^6$
11	Na	3	0	0	$-\frac{1}{2}$	$1s^2 2s^2 2p^6 3s^1$

Table 2.1: *Electronic configuration of the first 11 elements of the periodic table.*

relationships of the first 20 elements, after which it breaks down. Despite countless suggestions and efforts to coerce the theoretical Aufbau sequence into register with chemically observed periodicity, it has not been possible to define anything but a formal resemblance between the sequences.

Two other, partially successful models to account for elemental periodicity, were proposed before and forgotten after the advent of quantum theory. An anonymous proposal, later ascribed to Prout, was based on the assumption that all atoms are composites of hydrogen. The purpose of this proposal was to account for the statistically improbable distribution of relative atomic weights, close to integer values. Following the discovery of isotopes Prout's hypothesis gained some new respectability, but it has never been fully exploited. Another theory was summarized by its author [20] in the statement:

The properties of the elements are the properties of numbers.

Each of the three theories accounts for some, but not all aspects of elemental periodicity. The common ground among the three may well reveal the suspected link with space-time structure. What is required is to combine aspects of the wave-mechanical model of hydrogen, the structure of atomic nuclei and number theory.

2.2 Number Theory

One number theoretic concept with an obvious link to geometry and to the topology of space-time is the golden ratio, obtained as a root of the quadraric equation

$$x^2 - x - 1 = 0$$

By convention the two roots are identified by the symbols

$$\Phi = 1.61803398\ldots \quad \text{or} \quad \tau = 0.61803398\ldots$$

Some of the intriguing properties of the golden ratio include (i) its connection with self-similar structures, evidenced by its definition as a continued fraction:

$$x = 1 + \cfrac{1}{1 + \cfrac{1}{1 + \cfrac{1}{1 + \cfrac{1}{1 + \cdots}}}} = 1 + \frac{1}{x}, \quad i.e. \ x = \Phi$$

(ii) its appearance as the limit of the Fibonacci series that occurs in an amazing variety of biological and astronomical growth structures; and (iii) in defining the shapes of importance in aperiodic tilings.

The other concept from number theory of possible importance in the analysis of periodic functions is the distribution of prime numbers. This problem is still unsolved. The simplest of the known prime-number formulae, $p = 6n \pm 1$, where n is a natural number, rather than identify prime numbers, simply excludes all multiples of 2 and 3 (two-thirds of all natural numbers), from $\mathbb{N}$. With the complete set $\mathbb{N}$ arranged on a spiral with period 24, all prime numbers therefore appear on a conspicuous eight-arm cross.

2.2.1 Number Spiral and Periodicity

The prime number cross is shown in figure 1. Because of another property

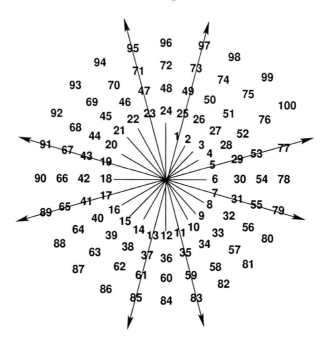

Figure 2.1: *Natural numbers arranged on a spiral of period 24*

of the number spiral the eight-line cross resembles the periodicity of the chemical elements. The property in question relates to the sums

$$\sigma_n = \sum_{i=24n}^{24(n+1)} i = (2n+1)a; \quad n = 0, 1, 2 \ldots; \quad a = 300$$

over all numbers along the spiral. The coefficients of a match the degeneracy of angular momentum wave functions of the hydrogen electron (T 5.2.1); each sum is therefore interpreted to yield the number of s, p, d, f, etc. electron pairs over a fixed number of atoms. The number $a = 300$ clearly exceeds the

group 1, and so forth.

To establish how the periodicity of the stable nuclides relates to that of the elements, the 264 stable isotopes, arranged in order of increasing mass number[1] are separated into 11 groups of 24 by inserting suitable straight lines in figure 2. These period lines are not parallel and they intersect each horizontal line that represents a fixed value of the ratio $Z/(A - Z)$ at a unique set of atomic numbers. It is found by inspection that the points of intersection with the line representing the ratio τ are in all respects identical to the period 2 and 8 atomic numbers of the table in figure 3. Another

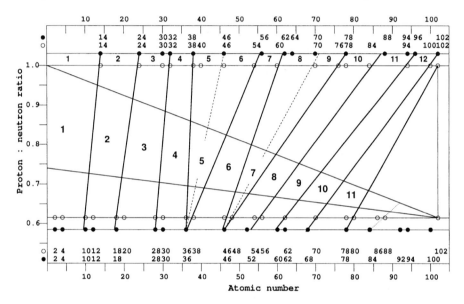

Figure 2.4: *Diagram to identify elemental periodicity in four special forms*

ratio of obvious interest is 1. Atomic numbers generated by intersection with this line show a remarkable inverse relationship with those on the τ line. In particular, pairs of numbers from the two sets, in significantly many cases, add up to 94. Some inversion symmetry between the sets is clearly at work, but equally obviously this symmetry is broken, or hidden. To reveal the full symmetry of the classification one must extrapolate further the period lines over the same distance in both directions, to produce the situation shown in figure 4. The appropriate extrapolation distance is $2\tau - 1.2 = .036..$ along

---

[1] Nuclides with a common mass number are arranged in order of increasing atomic number.

the vertical axis, whereby symmetrical arrangements occur at ratios of 1.036 and 0.582. The points at which the period lines intersect the 0.58 ratio line correspond exactly with atomic numbers at which energy levels, derived from the Schrödinger solution of the H atom are closed in the Aufbau procedure. The points of intersection at 1.04 are inverse to the Schrödinger solution in the sense that $4f < 3d < 2p < 1s$, *etc.* In addition these numbers are distributed symmetrically about 51.

The four periodic laws defined by figure 4 are related in the sense that each of them fits the compact periodic table (figure 3) such that all energy shells close in either period 2 or 8 [21]. From the spacing of points at the ratio 1.04 it is inferred that two extra groups of 24 nuclides become stable against β-type decay. The total number of nuclides is thereby increased to $12 \times 25 = 300$, as required by the number spiral. The total number of elements increases to 102-2=100 as required.

2.4   Atomic and Nuclear Structure

The stable chemical elements and their isotopes independently obey the same periodic law; an implication is that the extranuclear electronic structure of an atom is symmetry related to the structure of its nucleus. If such self-similarity between electronic configuration and nuclear structure indeed exists, it is necessary that intercepts of the period lines (figure 4) at $Z/(A-Z) = 0$ should correspond with the energy spectrum of the nucleus. After extrapolation and symmetry reflection across $Z = 51$, the diagram shown in figure 5 is obtained. All lines of figure 4 are shown suitably extrapolated, together with their symmetry related counterparts, where these also link up with periodic points of the six special arrangements. Only two lines, thinly drawn, fail to follow the pattern. The symmetric disposition around $Z = 51$ served as a guide to connect additional points at ratio 1.04 to associated points at zero ratio. On incorporating the symmetry-generated points at zero with those (< 51) obtained by extrapolation, a one-particle nuclear spectrum is extended to 102. Although this is not the observed shell-model spectrum, the latter appears in two segments at ratios 0.22 and 0.18.

The energy spectrum of the nucleus according to the semi-empirical shell model [23] appears not at zero ratio, but in two disjoint parts at ratios 0.22 and 0.18. This shift relates to the appearance of the symmetric arrangement at ratio 1.04 rather than 1. An Aufbau procedure based on this result fits the 8-period table derived from the number spiral, but like the observed periodic table, at ratio τ, the shell-model result also has hidden symmetry. At ratio zero, the inferred energy spectrum not only fits the 8-period table but also

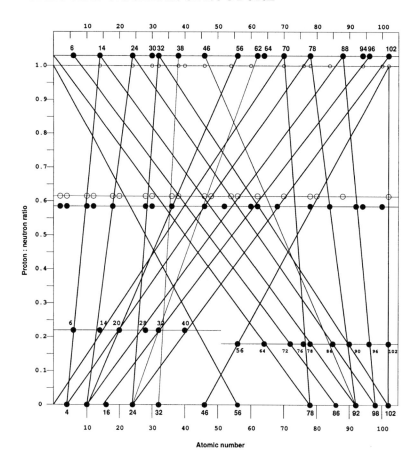

Figure 2.5: *Period lines of figure 3, together with those generated by the symmetry operator detected at the ratio 1.04, all extrapolated to zero ratio.*

exhibits the symmetry observed at ratio 1.04. The extra symmetry-generated lines shown in figure 5, are suggested to represent atoms of antimatter. This interpretation is rationalized on mapping the diagram on the double cover of a Möbius strip, shown in figure 6.

2.4.1 Physical Interpretation

The physical interpretation of the periodic relationship suggested by numerical pattern generation relies on the known effect of isotropic compression on the electronic structure of atoms [24]. Compression causes all energy levels to rise and removes the degeneracy of sub-levels. The effect becomes more pronounced with increasing quantum number l. Relative energies for hydro-

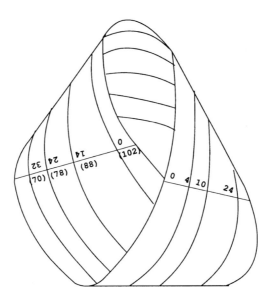

Figure 2.6: *Period lines of figure 5 mapped onto the double cover of a Möbius strip*

gen compressed to a radius of $0.1a_0$ have been calculated [25] to increase in an order

$$1s < 2p < 3d < 2s < 4f < 3p < 4d < 3s < 5f < 4p....$$

This sequence approaches the inversion inferred above. Hence the degrees of inversion represented by the spectra indicated at ratios 1.0 and 1.04, require enormous pressures that can be generated only in massive stars. To understand what happens in such a star we note that a pair of neighbouring points on any festoon mapped in figure 2 differs by the equivalent of an α-particle, with a proton:neutron ratio unity. A logical picture that relates to nuclear synthesis emerges.

Pressure within stars can increase to the point where sub-atomic particles fuse to form ^{2}H, ^{3}He, ^{4}He, and ^{5}He. The α-particle is the most stable of these units and becomes formed in sufficient excess to add progressively to each of four starting units to produce nuclides in four series of mass number $4n$, $4n \pm 1$, $4n - 2$, as observed [21]. Under these conditions the proton:neutron ratio for each series approaches unity with increasing mass number. At a certain age, a star of such magnitude explodes as a supernova to release the synthesized material into low-pressure environments in which a phase transition ensues. This transition consists of an inversion of energy levels and the radioactive decay of certain nuclides. Of 300 in total synthesized in the star, only 264 survive as stable nuclides in the solar system.

Conditions of pressure approaching infinite might occur in a black hole. Not only is the electronic spectrum inverted, but the electrons might be imagined to become pushed through the nucleus to emerge as positrons, leaving antiprotons behind. The matter that emerges with inverted chirality on the backside of the black hole rearranges to atoms of antimatter.

An inverse relationship between matter and antimatter [26, 11] provides a hint to explain the periodic tables identified at ratios τ and $1.2 - \tau$. The curvature of space at the singularity on the Schwarzschild radius in a black hole is infinitely high [26]. Elemental synthesis is postulated to occur at large, but finite, curvature. Schrödinger's equation, which describes a single electron in the field of an isolated proton, may be considered to define an empty universe with zero curvature. In the real world curvature is assumed small, but finite. All of these situations are mapped onto the Möbius surface. At ratio 1.04 atomic inversion is indicated by the symmetry that restricts the maximum atomic number of stable species to 102. Continuation in the realm of antimatter ends at zero ratio, on the upper end of the mass scale. This side reflects into the small-mass region of matter atoms, continuing to the full range of atomic numbers at 1.04.

2.4.2 The Number Pattern

The proposed interconversion between matter and antimatter is not inconsistent with a number system that assumes a chiral relationship between natural numbers (n) and their conjugates (in), $i = \sqrt{-1}$. This relationship is shown schematically in figure 7. The two sets of numbers project into the complex plane as an achiral set. The two positive spirals, shown emanating from the complex plane have opposite chirality. Should these spirals coalesce at infinity, an involution is implied. The resulting number pattern can be mapped also on the double cover of a Möbius strip as shown in figure 8. Shaded regions represent the surface on the reverse side of the unshaded region. Numbers represented by the two positive (or negative) spirals are involuted along the cover and merge with the conjugate spiral after transplantation of $\pi/2$. After displacement π each type is converted to its conjugate. At $2\pi/3$ signs have changed. The stippled lines show the formation of the Möbius strip. The involution serves to position conjugate pairs of chiral numbers in the surface, on opposite sides of an achiral interface, which represents the complex plane. Hence any number might be converted into its complex conjugate by either penetrating the interface (the complex plane) or following the number spiral in the surface, through a Möbius involution.

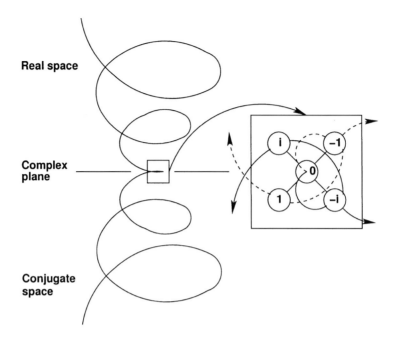

Figure 2.7: *Schematic diagram to represent natural numbers and their conjugates as two spirals that meet at infinity. The mirror image of each spiral represents negative numbers. Real and conjugate number spirals are chiral. In projection on orthogonal axes in the complex plane, together they create an achiral interface*

In the physical picture of nuclides and antinuclides this involution becomes identified with the symmetry inversion at zero ratio and penetration of the interface as passage through a black hole.

2.4.3   The Farey Sequence

The general number pattern that describes the proton:neutron ratio of stable nuclides has been identified. To understand the origin of this pattern in pure number theory it is noted that all possible ratios $A/(A-Z)$ must be rational fractions. The pattern of interest must hence be embedded in the algorithm that allows the ordering and enumeration of rational fractions, known as the Farey sequence [27]. This sequence is most readily generated as an inverted tree structure, by forming new fractions between $0/1$ and $1/1$ through separate addition of numerators and denominators, as shown in figure 9. An ordered array of rational fractions $Z/(A-Z)$ must enumerate all allowed proton:neutron ratios, such that the neutron excess $A-2Z \geq 0$. By defin-

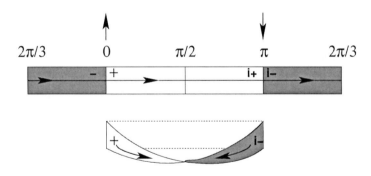

Figure 2.8: *The relationship between the four spirals of figure 7 is best understood by mapping on the double cover of a Möbius strip, opened up in this diagram.*

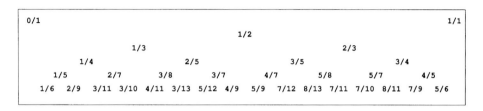

Figure 2.9: *Farey sequence of rational fractions. Starting with the first row and reading from right to left 3/7 is identified as the 12th rational fraction.*

ing A in modular form, *e.g.* $A(\text{mod}4)$, allowed compositions are disjoined into simpler, related subgroups that may be examined separately. In the simplest case $A(\text{mod})4 \equiv 0$, the allowed compositions for even Z correspond exactly with the $4m$ series of nuclides in their region of stability and trace out the same festoons $A - 2Z = 4n$ in the plot of Z vs $Z/(A - Z)$, as the stable nuclides, as shown in figure 10. To improve visual appreciation of the pattern only those fractions of rank (defined as the difference, denominator-numerator) 1 and 2 are shown. In a more detailed analysis it is found that the fractions representing allowed nuclides occur along the festoons in an order which is closely defined by the Farey sequence. Once it is known how to cut the region of stability from the infinite set of fractions, the natural occurrence of stable nuclides may be considered solved.

The stability problem is solved on noting that allowed fractions at small atomic number begin at unity and approach τ with increasing Z. This trend should, by definition follow Farey fractions determined by Fibonacci numbers. The first few Fibonacci numbers are 0,1,1,2,3,5,8,13,21, *etc.* The ra-

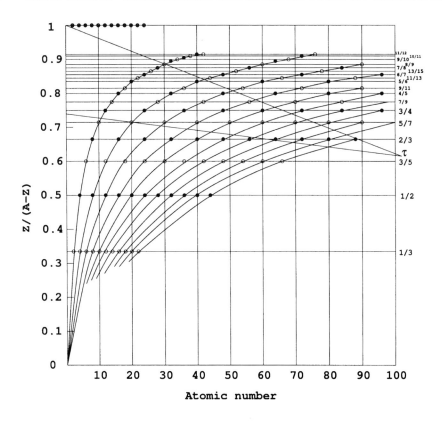

Figure 2.10: *Possible combinations of integers Z and $A-Z$ such that $A = 4m$ and $Z = 2n$, represented as rational fractions plotted against Z*

tional fractions that occur between the first non-zero Fibonacci fractions, *i.e.* 1/1 and 2/3 can be established by the Farey procedure. The resulting fractions are shown in figure 11. These fractions converge to 1/1 at the one end and 2/3 at the other end. On choosing 5/8 as the second limit the convergence proceeds in exactly the same way through 2/3 towards 5/8, and beyond towards τ, for higher Fibonacci fractions. According to figure 12 all relevant points of intersection on the straight line from 1 to τ that can refer to actual nuclides are generated by the Farey sequence that approaches 2/3. The value τ is approached at $Z = 102$.

All fractions beyond 3/4, at the centre of the Farey subset, are further seen to define points of intersection between the curves of constant $A - 2Z$ and the straight line between coordinates of (14/19,0) and (2/3,87), at the intersection with the line $1 \rightarrow \tau$. These lines provide the correct stability limits. They have the additional merit of automatically limiting the maximum allowed atomic number of a stable element to 83.

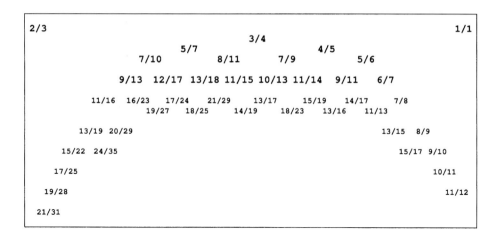

Figure 2.11: *Low-rank fractions in the Farey sequence between the Fibonacci fractions 1/1 and 2/3*

2.4.4   A Golden Parabola

Because the range of nuclidic stability is bounded by fractions that derive from Fibonacci numbers, it probably means that nuclear stability relates directly to the golden mean. To demonstrate this relationship it is noted that the plot of A vs Z, shown in figure 13 for the $A(\mathrm{mod}4) \equiv 0$ series of nuclides, separates into linear sections of constant neutron excess $(A - 2Z)$ and slope 2. Each section terminates at both ends in a radioactive nuclide. The range of stability for each section follows directly from

$$\begin{aligned} x^2 &= (\Delta A/2)^2 + (\Delta A)^2 \\ x &= (\sqrt{5}/2)\Delta A = (\Phi - 1/2)\Delta A \end{aligned}$$

as a function of the golden mean. The exact functional relationship may be derived from the definition of the golden mean by the more general expression, in terms of any integer n: $x^2 - x - n = 0$. This function defines the parabola shown as an inset in figure 14; it has a minimum $n = -\frac{1}{4}$ at $x = \frac{1}{2}$. A related parabola is obtained from the end members of the linear sections described in figure 13 on plotting the maximum and the negative of the minimum atomic number for each section, against neutron excess. The two parabolas coincide after scaling the theoretical x-coordinate by a factor $2\tau = 1.236$ and matching values of neutron excess to n. The minimum point of the curve shifts to τ. The relationship between the two sets of axes is

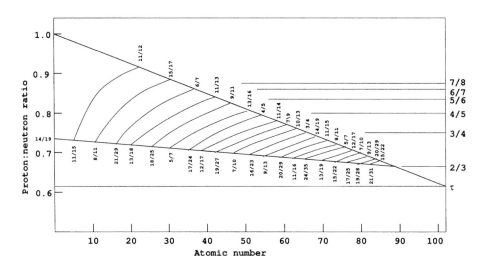

Figure 2.12: *Diagram to show points of intersection between curves of constant $A - 2Z$ and the straight lines that limit the occurrence of stable nuclides*

defined by $A - 2Z = 4n$. The resulting curve is described by the equation

$$x^2 - 2\tau x - 4\tau^2 n = 0$$

with solutions $x = \tau(1 \pm \sqrt{1 + 4n})$. The relationship between x and atomic number, after scaling is $Z = 10(2x - 1)$. These results establish the elusive general relationship that exists between neutron excess and atomic number. The scatter around the mean curve arises from two factors - the relatively lower stability of nuclides with odd mass number and the restriction of both Z and A to integer values.

2.4.5   Periodic Symmetry

To obtain the golden parabola (figure 14) that fits the stability bounds of nuclides it is necessary to scale the x-coordinate by a factor 2τ. Since the point $(x, n) = 0$ is not affected by the scaling, the resulting parabola is no longer centred at $x = \frac{1}{2}$, $(Z = 0)$, but at $x = \tau$, i.e. $Z = 10(1.236 - 1) = 2.36$. Shift of the golden parabola has an effect on the way in which a periodic function can be used to simulate the symmetrical periodic state of elemental electronic configuration identified in figure 6 at the ratio 1.04. Using a sine curve it is necessary to define

$$z = \sin 2\pi \left(\frac{i - 3}{32} \right) \quad , \quad i = 0, 102 \tag{2.1}$$

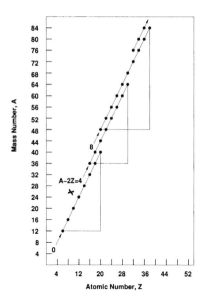

Figure 2.13: *Relationship between A and Z for nuclides of the series* $A\,(mod4) \equiv 0$

shown graphically in figure 15. Open and filled circles identify the completion of either f and p or d and s sub-shells in periods 8 or 2 respectively. This arrangement is the symmetrical periodic state that Reynolds [22] unsuccessfully tried to model in terms of vibrations on a stretched string.

It is assumed that the curve in figure 15 represents the maximum possible number of stable elements, under any conditions. Because atomic numbers 43 and 61 occur in no sequences generated by α-particle synthesis, this number becomes decreased to 100. Outside of massive stars not all elements are stable and the symmetry of the periodic function (eq. 1) is hidden. Such a situation is shown in figure 3, in which it is necessary to distinguish between atomic numbers (1-83) and symmetry numbers (1-102), and also in figure 16.

Relative to the symmetrical situation of figure 15, the number of stable elements is reduced to 81 and the energy-level sequence that dictates the Aufbau procedure is inverted. The structure that positions closed-shell configurations in periods 2 and 8 is maintained by the appearance of three gaps along the number spiral.

Symmetry remains hidden in all other situations that define known energy-level distributions, except the nuclear arrangement at zero ratio, for which full symmetry is displayed.

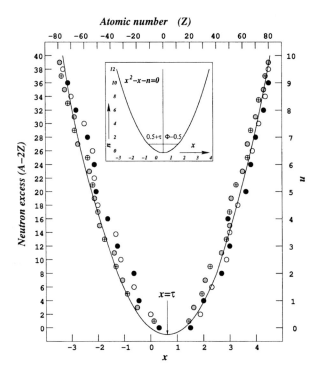

Figure 2.14: *Nuclides have a limited range of stability for each value of neutron excess, A-2Z. The maximum Z of these ranges plotted against A-2Z define the right arm of the parabola shown in the diagram. The left arm of the parabola is obtained by plotting minimum Z as negative integers. The curve is described by the equation* $x^2 - 2\tau x - 4\tau^2 n = 0$, *i.e.* $x = \tau(1 \pm \sqrt{1 + 4n})$

2.5   Space-Time Structure

The persistent correlation that recurs between number patterns and physical structures indicates a similarity between the structure of space-time and number. Like numbers and chiral growth, matter has a symmetry-related conjugate counterpart. The mystery about this antimatter is its whereabouts in the universe. By analogy with numbers, the two chiral forms of fermionic matter may be located on opposite sides of an achiral bosonic interface. In the case of numbers this interface is the complex plane, in the physical world it is the vacuum. An equivalent mapping has classical worlds located in the two surfaces and the quantum world, which requires complex formulation, in the interface.

Like numbers and their conjugates, matter and antimatter would merge naturally if the two conjugate surfaces constituted the double cover of a

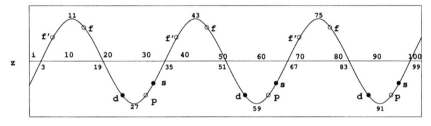

Figure 2.15: *Simulation of elemental periodicity by a sine curve. The periodic part occurs between atomic numbers 3 and 99*

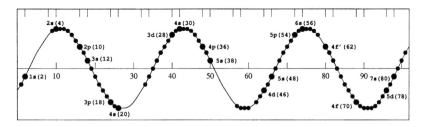

Figure 2.16: *Atomic positions superimposed on the sine wave that describes the hidden symmetry of periodic classification. Short vertical lines define allowed positions for closed electronic sub-shells; the horizontal axis represents symmetry numbers. Atomic numbers are shown in parentheses.*

Möbius strip. In this way the antimatter mystery disappears: matter and antimatter are one and the same thing, which merely appear to be different depending on their position in the double cover. In more dimensions the Möbius model is replaced by a projective plane, obtained from an open hemisphere on identifying points on opposite sides of the circular edge. Topologically equivalent constructs are known as a Roman surface or a Klein bottle.

This space-time model is a conjecture that has been described in detail [28] and will be reconsidered in chapter 7. A new aspect thereof, which derives from number theory, is that the general curvature of this space-time manifold [26, 29] relates to the golden mean. This postulate is required to rationalize the self-similar growth pattern that occurs at many levels throughout the observable universe.

Chapter 3

Bohmian Mechanics

Bohmian mechanics refers to the ontological interpretation of quantum theory pioneered by Bohm [4]. The mathematical structure of the theory is not affected by the different interpretation and the same formalism adopted before [7] will be used here.

3.1 Historical Introduction

Quantum theory was developed primarily to find an explanation for the stability of atomic matter, specifically the planetary model of the hydrogen atom. In the Schrödinger formulation the correct equation was obtained by recognizing the wave-like properties of an electron. The first derivation by Schrödinger [30] was done by analogy with the relationship that was known to exist between wave optics and geometrical optics in the limit where the index of refraction, n does not change appreciably over distances of order λ. This condition leads to the eikonal equation (T3.15)

$$(\nabla \phi)^2 = n^2$$

which resembles the Hamilton-Jacobi equation (T3.5)

$$(\nabla S)^2 = 2m(E - V)$$

that relates Hamilton's principal action function, S to the kinetic energy. The dimensionless quantity $2m(E - V)/p_0^2$, where p_0 has the dimensions of momentum, can be viewed as the mechanical analogue of n^2. It can be argued that a wave-mechanical equivalent of the action function should emerge where the potential energy suffers a large fractional change over the dimensions of the particle. A new function, analogous to the scalar potential

ϕ, of geometrical optics

$$\phi = \phi_0 \exp[i\{k \cdot r - \omega t\}]$$

is now defined through

$$\Psi = \psi_0 \exp[ik_0 S(r, p, t)]$$

or, in terms of Hamilton's characteristic function, W

$$\Psi = \psi_0 \exp\left[\frac{2\pi i}{\lambda_0}\left(W(r, p) - Et\right)\right] \tag{3.1}$$

to describe the behaviour of a particle in wave formalism. Evidently

$$\frac{W(p, r)}{\lambda_0} \propto \frac{k \cdot r}{2\pi} \text{ and } E \propto \frac{\lambda_0 \omega}{2\pi}.$$

To satisfy the conditions of wave mechanics the relationships are written in the form

$$p = \hbar k \tag{3.2}$$
$$E = \hbar \omega \tag{3.3}$$

defining the quantum conditions that lead to Schrödinger's equation. With these substitutions eqn (1) reduces to

$$\Psi = \psi_0 \exp\left[\frac{2\pi i}{h}(pr - Et)\right]$$

The operator representation for linear momentum and energy is derived from the derivatives

$$\frac{\partial \Psi}{\partial r} = \Psi\left(\frac{2\pi p i}{h}\right), \text{ i.e. } p \to \frac{h}{2\pi i}\frac{\partial}{\partial r}$$

and

$$\frac{\partial \Psi}{\partial t} = \Psi\left(-\frac{2\pi i E}{h}\right), \quad E \to -\frac{h}{2\pi i}\frac{\partial}{\partial t}$$

Hence the Hamiltonian $H = \frac{p^2}{2m} + V$ goes into

$$-\frac{h}{2\pi i}\frac{\partial \Psi}{\partial t} = \left(-\frac{h^2}{8\pi^2 m}\frac{\partial^2}{\partial r^2} + V\right)\Psi$$

which is Schrödinger's equation

$$i\hbar\frac{\partial \Psi}{\partial t} = \left(-\frac{\hbar^2}{2m}\nabla^2 + V\right)\Psi \tag{3.4}$$

The contours along which Hamilton's principal function S remains constant describe moving surfaces in Euclidean space. The gradient ∇S at each point of a moving surface is orthogonal to the surface, and the particle trajectories associated with S are given by the solutions to

$$m\dot{\boldsymbol{x}} = \nabla S$$

It follows that a family of trajectories may be obtained by constructing the normals to S, each one distinguished by its starting point $\boldsymbol{x}_o$. This description of particle motion is equivalent to the classic Huygens construction of elementary waves and their envelopes, with mechanical action taking the role of the phase in wave formalism.

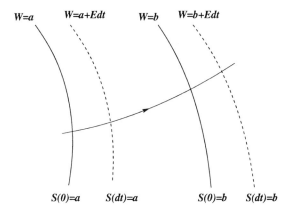

Figure 3.1: *The motion of surfaces of constant S in configuration space. At $t = 0$ the surfaces $S = a$ and $S = b$ coincide with the surfaces for which $W = a$ and $W = b$. Surfaces of constant W have fixed locations in space. Surfaces of constant S represent wavefronts propagating in configuration space. The trajectory of a single particle in 3D-space lies along the wave normals.*

It appears that once Schrödinger's equation had been found the original quest to explain the stability of matter receded into the background. Maybe it was agreed that the appearance of quantized stationary states was an adequate explanation of atomic stability. In fact however, Schrödinger's equation offers no more than a mathematical description of electronic bound states and certainly not a physical model to account for the stability of the observed arrangement. Proposals by two of the pioneers, De Broglie and Madelung, of a physical basis for wave mechanics generated little enthusiasm. The hydrodynamic model proposed by Madelung [31], as well as De

the quantum limit with classical trajectories and the dynamical properties of individual particles.

Bohm [4] demonstrated that the close parallel of the classical Hamilton-Jacobi (HJ) equation (T3.4) with Schrödinger's equation provides a logical point of departure for a causal account of quantum events. A wave function in polar form with real amplitude R and phase S,

$$\Psi = R \exp\left(\frac{2\pi i S}{h}\right)$$

with derivatives

$$\frac{\partial^2 \Psi}{\partial x^2} = \left[\frac{\partial^2 R}{\partial x^2} + \frac{2i}{\hbar}\frac{\partial R}{\partial x}\frac{\partial S}{\partial x} + \frac{Ri}{\hbar}\frac{\partial^2 S}{\partial x^2} - \frac{R}{\hbar^2}\left(\frac{\partial S}{\partial x}\right)^2\right] e^{iS/\hbar}$$

$$\frac{\partial \Psi}{\partial t} = \left(\frac{\partial R}{\partial t} + \frac{iR}{\hbar}\frac{\partial S}{\partial t}\right) e^{iS/\hbar}$$

is substituted into Schrödinger's equation (4) and multiplied throughout by $\exp(-iS/\hbar)$ to give

$$i\hbar\frac{\partial R}{\partial t} - R\frac{\partial S}{\partial t} = -\frac{\hbar^2}{2m}\frac{\partial^2 R}{\partial x^2} + \frac{R}{2m}\left(\frac{\partial S}{\partial x}\right)^2 + VR - \frac{\hbar i}{2m}\left[R\frac{\partial^2 S}{\partial x^2} + 2\frac{\partial R}{\partial x}\frac{\partial S}{\partial x}\right]$$

After separation of the real and imaginary parts two equations are obtained

$$\frac{\partial R}{\partial t} = -\frac{1}{2m}\left[R\frac{\partial^2 S}{\partial x^2} + 2\frac{\partial R}{\partial x}\frac{\partial S}{\partial x}\right]$$

$$-\frac{\hbar^2}{2m}\frac{\partial^2 R}{\partial x^2} + \frac{R}{2m}\left(\frac{\partial S}{\partial x}\right)^2 + VR + R\frac{\partial S}{\partial t} = 0$$

The first of these equations rearranges into

$$\frac{\partial}{\partial t}\left(R^2\right) + \frac{\partial}{\partial x}\left(\frac{R^2}{m}\frac{\partial S}{\partial x}\right) = 0 \tag{3.5}$$

and the second into

$$\frac{\partial S}{\partial t} + \frac{1}{2m}\left(\frac{\partial S}{\partial x}\right)^2 - \frac{\hbar^2}{2mR}\frac{\partial^2 R}{\partial x^2} + V = 0$$

In three-dimensional form it reads

$$\frac{\partial S}{\partial t} + \frac{(\nabla S)^2}{2m} - \frac{\hbar^2}{2mR}\nabla^2 R + V = 0 \tag{3.6}$$

Equation (5) is the analogue of the classical continuity equation

$$\frac{\partial \rho}{\partial t} + \nabla(\rho \boldsymbol{v}) = 0$$

and serves to identify quantum-mechanical probability density with $P = R^2$ and momentum with $\boldsymbol{p} = \nabla S$. Equation (6) is identical with the classical HJ equation, except for the term

$$V_q = -\frac{h^2 \nabla^2 R}{8\pi^2 m R}$$

called the quantum potential, which represents the only difference between the classical and non-classical equations. Quantum behaviour is therefore evident when V_q has an appreciable value, which happens when m is small. Alternatively the classical limit occurs in the limit $(h/m) \to 0$, *i.e.* for massive objects.

It is noted that a system of particles reaches equilibrium when the resulting forces on them are zero, and hence the quantum force on a free particle must be perceived to vanish. This requires the quantum potential to be either zero or a constant, independent of position. The first condition relates to a classical particle, whereas the second condition implies

$$\nabla^2 R \pm \frac{8\pi^2 m R k}{h^2} = 0$$

which defines R as either an exponential or an oscillatory function, i.e. the quantum condition. Note that $R \to \psi$ for $k = T$, the kinetic energy, to produce the time-independent Schrödinger amplitude equation:

$$\nabla^2 \psi = \frac{8\pi^2 m}{h^2}(E - V)\psi = 0 \tag{3.7}$$

3.3 The Ontological Interpretation

In terms of the quantum-potential formulation particle trajectories can be associated with the quantum HJ equation (6) in exactly the same way as in the classical case [34, 35]. As before, particle trajectories associated with the phase S may be obtained by constructing the normals to S, each one distinguished by its initial coordinates. By this procedure Bohm managed to revive the pilot-wave model of De Broglie. It means that a point particle of mass m on a trajectory $\boldsymbol{x} = \boldsymbol{x}(t)$, is now associated with the physical wave Ψ propagating in space. This interpretation allows an electron to be described with a well-defined position $\boldsymbol{x}(t)$ which varies continuously and is causally determined.

The particle is now separate from the quantum field that affects it. This field is given by

$$\Psi = R \exp\left(\frac{iS}{\hbar}\right).$$

Although Ψ is a field as real as Maxwell's fields, it does not show up immediately as the result of a single measurement, but only in the statistics of many measurements. It is the De Broglie-Bohm variable $\boldsymbol{x}$ that shows up immediately each time. The particle has an equation of motion (T3.2)

$$m\ddot{x} = -\nabla V - \nabla V_q$$

This equation shows that in addition to the classical force $-\nabla V$ acting on the particle there is also the quantum force $-\nabla V_q$.

The particle momentum is restricted to $\boldsymbol{p} = \nabla S$. Since the wavefunction Ψ is single-valued it follows that R must be a single-valued function of position. However, the phase function is not uniquely fixed and two S functions like S' and S, where

$$S' = S + nh$$

give rise to the same Ψ for integer n. On traversing a closed loop in space it is therefore possible for the phase to change by an amount

$$\oint dS = \oint \nabla S dx = nh$$

This equation resembles the Bohr-Sommerfeld condition $\oint p\,dx = nh$, but differs from it in that S now is the solution of the quantum HJ equation and not of the classical one as before.

The intensity of a wave Ψ is proportional to the square of the amplitude, i.e. $I = R^2$. Multiplication of the amplitude by a real constant therefore scales the intensity, but the quantum potential stays the same,

$$V_q = -\frac{\hbar^2}{2m} \frac{\nabla^2(aR)}{aR} = \frac{\hbar^2}{2m} \frac{\nabla^2 R}{R}$$

The effect of V_q is seen to be independent of the intensity of the quantum field and to depend only on its form. This is in sharp contrast to the effect of classical waves. The effect of the quantum potential on a quantum particle has been likened to a ship on automatic pilot being guided by radio waves. Here, too, the effect of the radio waves is independent of their intensity and depends only on their form. The ship moves with its own energy, and the form of the radio waves is taken up to direct the much greater energy of the ship. This implies that a particle moving in empty space and the absence of classical forces need not travel uniformly in straight lines. This is a radical non-classical departure from Newtonian theory. Moreover, since the effect of the waves does not necessarily fall off with the distance, even remote features of the environment can profoundly affect the movement.

In a double-slit experiment it is proposed that the particle goes through one of the slits but the wave goes through both, causing interference behind the slits and hence operating as a pilot wave that guides the particle into regions of constructive interference.

3.3.1 Hydrodynamic Analogy

The causal interpretation of quantum theory as proposed by De Broglie and Bohm is an extension of the hydrodynamic model originally proposed by Madelung and further developed by Takabayasi [36]. In Madelung's original proposal R^2 was interpreted as the density $\rho(x)$ of a continuous fluid with stream velocity $v = \nabla S/m$. Equation (5) then expresses conservation of fluid, while (6) determines changes of the velocity potential S in terms of the classical potential V, and the quantum potential

$$-\frac{\hbar^2}{2m}\frac{\nabla R}{R} = -\frac{\hbar^2}{4m}\left[\frac{\nabla^2\rho}{\rho} - \frac{1}{2}\left(\frac{\nabla\rho}{\rho}\right)^2\right]$$

The quantum potential therefore arises in the effects of an internal stress in the fluid and depends on derivatives of the fluid density rather than on external factors.

This model is not adequate in itself as it contains nothing to describe the actual location $x(t)$ of the particle which is required for a causal interpretation of quantum theory. It is therefore necessary to postulate a particle that takes the form of a highly localized inhomogeneity that moves with the local fluid velocity $v(x, t)$. The inhomogeneity could be of density close to that of the fluid, which is simply being carried along with the local velocity of the fluid. As in any macroscopic fluid random fluctuations are assumed [37] to occur in the Madelung fluid. It is shown that such fluctuations may lead to the statistical result, $P = |\Psi|^2$.

The one-particle model is readily extended to the case of many particles. The wave function $\psi(x_1, x_2, \ldots, x_N, t)$ is defined in $3N$-dimensional configuration space. On writing $\psi = R\exp(iS/\hbar)$ a set of $3N$ velocity fields,

3.4.1   The EPR Effect

Although Bohm formalism provides the most forceful demonstration of the non-local character of quantum theory, the evidence has been around for many decades. The so-called EPR effect was first recognized by Einstein and his co-workers, Podolsky and Rosen in 1935 [3]. The purpose of their work was to demonstrate that the apparent non-local nature of quantum mechanics could only mean that a vital element was missing from the theory. The missing element had to be such as to counteract the non-local feature.

The EPR effect was demonstrated in terms of a Gedanken experiment to show that quantum systems which became correlated at any time, would stay correlated until the combined system is disturbed by measurement. Two principles of fundamental importance are assumed in this analysis:

1. If, without in any way disturbing a system, the value of a physical quantity in that system can be predicted with certainty (with probability equal to one), then there exists an element of reality corresponding to the physical quantity.

2. Consider an arbitrary observable A with a set of eigenfunctions, ψ_a, belonging to a series of eigenvalues denoted by a. According to the first assumption there is an element of reality corresponding to observable A in the system. Next, consider another observable B which does not commute with A, so that there exists no wave function for which A and B have simultaneously definite values. If every element of physical reality must have a counterpart in a complete physical theory, the previous conclusion implies that A and B cannot exist simultaneously.

In other words, for a pair of observable properties characterized by non-commuting operators, $AB \neq BA$, precise knowledge of property A precludes any knowledge of B.

In the thought experiment two systems I and II are assumed to interact during the period $0 \leq t \leq T$, after which there is no further interaction. Suppose that the states of both systems are known at $t < 0$. The state of the combined system at any $t > 0$ can be calculated, in particular at $t > T$. However, the state of any one of the systems can no longer be calculated after the interaction. According to quantum mechanics such information can only be obtained by further measurement, that amounts to reduction of the wave packet.

The eigenvalues $a_1, a_2, \ldots$ of property A pertaining to system I belong to the eigenfunctions $u_1(x_1), u_2(x_1), \ldots$ *etc.*, where x_1 represents the variables that describe system I. After interaction $(t > T)$ the correlated system is described by the wave function

$$\Psi(x_1, x_2) = \sum_{n=1}^{\infty} \psi_n(x_2) u_n(x_1)$$

where x_2 are the variables that describe system II. Now suppose that A is measured at value a_k. The measurement leaves system I in the state $u_k(x_1)$, the second system in state $\psi_k(x_1)$, and the wave packet is reduced to $\psi_k(x_2)u_k(x_1)$. The set of functions $u_k(x_1)$ is determined by the option to measure quantity A. An alternative would be to measure property B, with eigenvalues $b_1, b_2 \ldots$, etc., and eigenfunctions $v_1(x_1), v_2(x_1) \ldots$. The wave function at $t > T$ is now written as

$$\Psi(x_1, x_2) = \sum_{s=1}^{\infty} \varphi_s(x_2) v_s(x_1)$$

and measurement of B gives the eigenvalue b_r, leaving the wave packet $\varphi_r(x_2)v_r(x_1)$. Two different measurements on system I therefore leaves system II in states with two different wave functions, $\psi_k(x_2)$ and $\varphi_r(x_2)$. Since the two systems no longer interact at the time of measurement, no real change in system II should be brought about by anything done to system I.

It may be that the wave functions are eigenfunctions of two non-commuting operators corresponding to physical quantities such as p (momentum) and q (position) respectively. Then, by measuring either A or B in system I, it becomes possible to predict with certainty and without disturbing the second system, either the value of p_k or q_r. In the first case p is an element of reality and in the second case q is an element of reality. But ψ_k and φ_r belong to the same reality. This conclusion contradicts the assertion that non-commuting operators cannot have simultaneous reality. It was inferred that quantum theory is incomplete.

The Bohm Example

Bohm [39] presented the same argument in terms of an executable experiment involving measurement of the non-commuting projections σ_x, σ_y, σ_z of spin. It starts with a diatomic molecule in a singlet state of total spin zero, each atom with spin $\hbar/2$. Now suppose that the molecule is disintegrated by some process that does not change the total angular momentum. The two atoms will begin to separate and will soon cease to interact appreciably. Their combined spin angular momentum, however, remains equal to zero, because by hypothesis, no torques have acted on the system.

The spin of atom A is now assumed measured in the z-direction and a given result is obtained. Because of the entanglement of the two particles the spin of atom B is predicted to be the opposite of the measured value. Since atom A is disturbed by the measurement it cannot be argued that its spin was defined beforehand as an independent element of reality. However, the spin of atom B can be predicted without disturbing that atom in any way, only from the measurement of σ_z for atom A. The z-component of spin of atom B must arguably have been an element of reality all along.

Instead of σ_z, any other component of the spin of atom A could have been measured to show that the corresponding spin component of atom B was an element of reality. In this way all components of spin of atom B can be demonstrated to have simultaneous reality. Once more, quantum theory appears to be conceptually incomplete.

The argument so far is premised on the assumption that there is no interaction possible between the correlated atoms at the time that the measurement is done. However, should there be an undisclosed interaction between the two particles, disturbance of A, caused by the measurement could communicate itself to B, causing a simultaneous disturbance at that position as well. This interaction may result in orientation of the spin at B in a direction opposite to that of A. Not surprisingly Einstein, the father of special relativity, did not even consider such a possibility since it requires instantaneous non-local interaction.

Non-locality in terms of special relativity is best explained by the Minkowski space-time diagram, shown in figure 2. A stationary object follows a world-

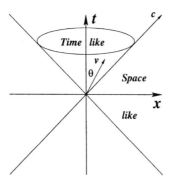

Figure 3.2: *Minkowski diagram showing three space dimensions* x *drawn perpendicular to the direction of time flow t.*

line in the direction of t. An object that moves with velocity v changes its position x as a function of time and traces out a world-line at an angle θ

with t. The maximum possible velocity in special relativity (T4.3.2) corresponds to c, the speed of light, which defines the limiting cone within which relativistic world-lines can occur. Events in the light cone are called time-like and those in the excluded region are called space-like. Space-like or non-local events imply that $v > c$.

3.4.2 Bell's Inequality

The mathematical formulation of the EPR effect in terms of which an experimental test of non-local interaction became possible was first presented by Bell [40] and further refined by Clauser and co-workers [41]. The crucial experiment was envisaged in terms of spin measurements by Stern-Gerlach magnets (T5.1.2) on two separated particles that constitute a singlet spin system. The two magnets had to be arranged so as to measure different selected components of the spins σ_1 and σ_2 at the two positions. In the absence of any interaction between the magnets or the spins, the degree of correlation between measurements can be worked out in terms of simple probability distributions. This result will be considered to represent the prediction of realistic local theory, to be compared to the degree of correlation predicted by quantum theory. The extent to which the two theories deviate from experiment can be assessed by Bell's inequality.

If measurement of the component $\sigma_1 \cdot a$, where a is some unit vector, yields the value $+1$ then, according to quantum mechanics, measurement of $\sigma_2 \cdot a$ must yield the value -1 and *vice versa*. Since the measurements are conducted at places remote from one another the orientation of one magnet does not influence the result obtained with the other. Since the result of measuring any chosen component of σ_2 can be predicted by first measuring the same component of σ_1, it follows that the result of any such measurement must actually be predetermined. The initial wave function does not determine the result of an individual measurement so that this predetermination implies the possibility of a more complete specification of the state, as inferred by EPR.

The more complete specification may be effected by means of parameters λ, that may remain unspecified and hidden. The result A of measuring $\sigma_1 \cdot a$ is then determined by a and λ, and the result B of measuring $\sigma_2 \cdot b$ by b and λ. It follows that

$$A(a, \lambda) = \pm 1 \quad ; \quad B(b, \lambda) = \pm 1 \qquad (3.10)$$

The vital assumption is that the result B for particle 2 does not depend on the setting a, of the magnet for particle 1, nor A on b.

If $\varrho(\lambda)$ is the probability distribution of λ then the expectation value of the product of the two components $\sigma_1 \cdot a$ and $\sigma_2 \cdot b$, according to local realistic theory is

$$P(a, b) = \int \varrho(\lambda) A(a, \lambda) B(b, \lambda) d\lambda \qquad (3.11)$$

This probability should equal the quantum-mechanical expectation value, which for the singlet state is

$$\langle \sigma_1 \cdot a \sigma_2 \cdot b \rangle = -a \cdot b \qquad (3.12)$$

It was shown by Bell that this proposition is not possible.

Proof   Because ϱ is a normalized probability distribution

$$\int \varrho(\lambda) d\lambda = 1$$

and because of the properties (10), P in (11) cannot be less than -1. It reaches -1 when $A(a, \lambda) = -B(a, \lambda)$. Assuming this, (11) can be rewritten

$$P(a, b) = - \int \varrho(\lambda) A(a, \lambda) A(b, \lambda) d\lambda$$

If c is another unit vector, it follows that

$$
\begin{aligned}
P(a, b) - P(a, c) &= - \int \varrho(\lambda)[A(a, \lambda) A(b, \lambda) - A(a, \lambda) A(c, \lambda)] d\lambda \\
&= \int \varrho(\lambda) A(a, \lambda) A(b, \lambda)[A(b, \lambda) A(c, \lambda) - 1] d\lambda
\end{aligned}
$$

using (10), whence

$$| P(a, b) - P(a, c) | \le \int \varrho(\lambda)[1 - A(b, \lambda) A(c, \lambda)] d\lambda$$

The second term on the right is $P(a, c)$, whence

$$| P(a, b) - P(a, c) | \le 1 + P(b, c)$$

On inserting a fourth unit vector d, the symmetrical form of Bell's inequality [41] is obtained from

$$| P(a, b) - P(a, c) | \le 2 \pm [P(d, c) + P(d, b)]$$

as

$$| P(ab) - P(ac) | + | P(dc) + P(db) | \le 2 \qquad (3.13)$$

This inequality must be satisfied for a local hidden variable theory to apply to the singlet system of two particles with spin. A test for locality on the basis of the measurement of four sets of correlations becomes possible in terms of the inequalities.

Example The prediction of quantum theory (12) is that

$$P(\mathbf{ab}) = -\mathbf{a} \cdot \mathbf{b}$$

the inner product of the polarization vectors shown in figure 3. The angles

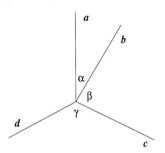

Figure 3.3: *Angular relation between the axes of various spin measurements.*

between the various polarization directions are defined by $\alpha = \mathbf{b} - \mathbf{a}$, $\beta = \mathbf{c} - \mathbf{b}$, and $\gamma = \mathbf{d} - \mathbf{c}$. Substituted into (13) the inequality becomes

$$I_B = \mid P(\alpha) - P(\alpha + \beta) \mid + \mid P(\gamma) + P(\beta + \gamma) \mid \leq 2$$

For $\alpha = \beta = 60°$ and $\gamma = 0$ the quantum inequality on the left, evaluated from the cosines is $5/2$, which violates the inequality. With three magnets placed 60° apart as in figure 3, measurements of the different spin components of a singlet system are therefore predicted by equations (12) and (13) to correlate differently if the interactions were quantum-mechanical or local, respectively. An outcome $I_B \leq 2$ would favour realistic local theory while $I_B > 2$ favours quantum theory. The EPR argument can hence be settled experimentally.

3.4.3 Entanglement

The idea of action at a distance was resisted both by Newton, and by Einstein [42] who called it "spooky", but it has now been demonstrated experimentally [43, 44] that local realistic theory cannot account for correlations between measurements performed at well separated sites. The conclusion is that quantum theory permits hidden variables and is non-local. This conclusion is at variance with relativity, but, as pointed out by Bohm [34], the non-locality of quantum theory only applies to complex wave functions and does not imply that the quantum potential can be used to transmit signals faster than light.

The first experiments to analyze EPR correlations used polarized light beams rather than electronic spin systems. The results obtained by Aspect [44] are especially relevant since the systems for study were prepared to be separated space-like. Aspect analyzed the polarization of pairs of photons emitted by a single source toward separate detectors. Measured independently, the polarization of each set of photons fluctuated in a seemingly random way. However, when two sets of measurements were compared, they displayed an agreement stronger than could be accounted for by any local realistic theory.

Because of the EPR effect quantum systems that have interacted before remain correlated even when the interaction no longer persists. The experiments have shown that, even when all interaction comes to an end, information about the second of a pair of particles can be obtained by performing a measurement on the first. The conclusion is that the physical world cannot be correctly described by a realistic local theory. It is necessary either to abandon the criterion of reality or to accept the possibility of action at a distance. The latter occurs because each particle is described by a wave function which is, in general, a non-local entity that collapses when a measurement is made. This collapse is instantaneous and complete. It occurs everywhere, also at the position of a particle not involved in the measurement and therefore predicts the correlation of distant events. Most particles or aggregates of particles, usually regarded as separate objects, have interacted in the past with other objects and must hence remain correlated and to constitute an indivisible entangled whole. This observation represents the scientific rediscovery [45] of holism [46].

One conceptual problem is to understand the existence of individual objects if non-local entanglement can spread across the entire universe. That would also prevent the practice of science which traditionally defines systems for study by isolating them, at least approximately, from their environment [47]. However, all sytems are not correlated equally well. Whenever a wave function can be written as a product

$$\psi(\boldsymbol{r}_1, \boldsymbol{r}_2, t) = \phi_A(\boldsymbol{r}_1, t)\phi_B(\boldsymbol{r}_2, t)$$

the quantum potential becomes the sum of two terms

$$V_q(\boldsymbol{r}_1, \boldsymbol{r}_2, t) = V_q^A(\boldsymbol{r}_1, t) + V_q^B(\boldsymbol{r}_2, t)$$

in which

$$V_q^A(\boldsymbol{r}_1, t) = -\frac{h^2}{8\pi^2 m}\frac{\nabla_1^2 R_A(\boldsymbol{r}_1, t)}{R_A(\boldsymbol{r}_1, t)}$$

and

$$V_q^B(\boldsymbol{r_2}, t) = -\frac{h^2}{8\pi^2 m} \frac{\nabla_2^2 R_B(\boldsymbol{r_2}, t)}{R_B(\boldsymbol{r_2}, t)}$$

In this case the two systems evidently behave independently. Situations like this are fairly common in chemistry, generally associated with an approach to the classical limit in which the quantum potential becomes negligible and non-local interactions insignificant. Although the basic law therefore refers inseparably to the whole universe, it tends to fragment into numerous independent parts, each constituted of further sub-units that are non-locally connected internally. The key to this fragmentation is the lack (or nature) of chemical interaction between sub-units, which can be treated in the traditional way.

Elucidation of the non-local holistic nature of quantum theory, first discerned by Einstein [3] and interpreted as a defect of the theory, is probably the most important feature of Bohm's interpretation. Two other major innovations that flow from the Bohm interpretation are a definition of particle trajectories directed by a pilot wave and the physical picture of a stationary state.

The trajectory description follows directly from a comparison of classical and quantum HJ equations, which merge as $V_q \to 0$. This can be interpreted to mean that the quantum electron has an unspecified diffuse structure that becomes concentrated when external factors force it into a confined space. In strictly classical terms the electron has radius

$$r = \frac{e^2}{4\pi \epsilon_o m_o c^2} = 2.8 \times 10^{-15} m$$

and a classical trajectory defined by the gradient to the moving surface represented by Hamilton's principal function S in eq. 6 for a particle of relativistic mass m and $V_q = 0$. There is no other distinction between a quantum-mechanical and a classical electron. The quantum-mechanical action S must therefore be interpreted exactly like its classical analogue as a surface that specifies possible trajectories of the electron. The actual trajectory of a given electron depends only on its starting coordinates $\boldsymbol{x_o}$. The De Broglie-Bohm model views the associated wave as a pilot wave that directs the motion of a particle, like a radio beam guiding a ship.

3.5 Stationary States

Stationary states are eigenfunctions of the Hamiltonian operator

$$H\Psi(\boldsymbol{x}, t) = E\Psi(\boldsymbol{x}, t)$$

$$H = -\frac{\hbar^2}{2m}\nabla^2 + V$$

The requirement that Ψ satisfy Schrödinger's equation fixes its temporal dependence

$$\Psi(\boldsymbol{x}, t) = \Psi_o(\boldsymbol{x})e^{\frac{-iEt}{\hbar}} \tag{3.14}$$

This solution is valid only if the external potential V is independent of time: $V = V(\boldsymbol{x})$. The initial function $\Psi_o = \psi$ is not arbitrary but must satisfy the time-independent Schrödinger equation. Writing $\Psi = R\exp(iS/\hbar)$ therefore gives

$$R(\boldsymbol{x}, t) = R_o(\boldsymbol{x})$$

$$S(\boldsymbol{x}, t) = S_o(\boldsymbol{x}) - Et$$

From this follows that

  (i) the probability density $\mid \Psi \mid^2 = R_o^2(\boldsymbol{x})$,

  (ii) the quantum potential is temporally independent, whence

$$\frac{\partial}{\partial t}(V + V_q) = 0$$

  (iii) if the wave function $\Psi_o \equiv \psi$ is real it implies that $S_o(\boldsymbol{x}) = 0$ and hence $\nabla S = \boldsymbol{p} = 0$ and $E = V + V_q$.

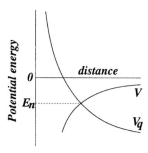

Figure 3.4: *The balance between Coulombic and quantum potentials in the s ground state of hydrogen*

The stationary solutions are eigenfunctions of the time-independent wave equation (7), characterized by constant V_q. For an atom in an s-state (or any Y_{nl0}-state) the wave function is real, which means that the electron is at rest. This result may seem surprising, because classically a dynamic equilibrium is advanced to explain why the potential does not cause the particle to fall into the nucleus. However, as $E = V + V_q$, one can argue that the quantum potential cancels the spatial variation of the classical potential to leave a constant energy E_n that is independent of position, as shown in figure 4.

$$\frac{\partial E}{\partial x} = \frac{\partial V}{\partial x} + \frac{\partial V_q}{\partial x} = 0$$

The acceleration of the particle by the quantum potential $-\nabla V_q/m$ balances the classical acceleration $\nabla V/m$ so that the particle remains in a fixed position and is prevented from falling into the nucleus by the outward acceleration due to the quantum potential.

For degenerate states more than one state with eigenvalue E take the form (9). Both are stationary, but a complex superposition, although again a stationary function, does not have the form (9) and could describe particles in motion [48].

3.6 Angular Momentum

Angular momentum, arguably the second most important concept in the understanding of chemistry is often either ignored or treated in such abstract mathematical terms that its true meaning remains obscure. This happens because the angular-momentum relationship, equivalent to the momentum and energy relationships, (2) and (3), that connect particle and wave properties of quantum-mechanical entities, is routinely overlooked [50]. The Planck-Einstein relationship (3) attributes a well-defined energy $E = \hbar\omega$ to any phenomenon with harmonic time dependence of periodicity τ, where $\omega = 2\pi/\tau$. The De Broglie relationship (2), likewise associates a well-defined momentum $p = \hbar k$ to a phenomenon with harmonic space variation of wavelength λ, where $k = 2\pi/\lambda$.

Table 3.1: *Quantum-mechanical particle-wave relationships.*

Invariance	Period	Pulsation	Dynamic	Quantum condition
Temporal	τ	$\omega = 2\pi/\tau$	E Energy	$E = \hbar\omega$ Planck-Einstein
Translation	λ	$k = 2\pi/\lambda$	p Momentum	$p = \hbar k$ De Broglie
Rotation	α	$m = 2\pi/\alpha$	L_z Angular momentum	$L_z = \hbar m$ Lévy-Leblond

It was suggested by Lévy-Leblond [50] that a well-defined component of angular momentum should accompany a phenomenon of periodicity α around a rotation axis along z, according to a relationship comparable to (2) and (3), *i.e.*

$$L_z = \hbar m \tag{3.15}$$

where $m = 2\pi/\alpha$. To ensure that the phenomenon repeats itself after a complete rotation through an angle of 2π it is required that α be a submultiple of 2π and m an integer. The argument, as summarized in table 1, is consistent with the conservation laws of momentum, energy and angular momentum that arise from the invariance of space-translation, time-translation and rotation, respectively.

Mathematically m derives as an integer from the requirement that the angular wave stays in phase with itself, *i.e.*

$$\exp(im\phi + 2\pi im) = \exp(im\phi)$$

$$\exp(2\pi im) = \cos 2\pi m + i \sin 2\pi m = 1$$

The angular wave number m may be either a positive or a negative integer depending on the sense of rotation of harmonic angular waves.

Now consider the components L_x, L_y and L_z of the angular momentum $\mathbf{L}$ along three orthogonal axes. For them to simultaneously take on unique and well-defined integer values m_x, m_y and m_z, the system should be in a state of rotational harmonicity around the three axes. This condition is impossible when dealing with travelling waves. Since stationary waves result from the superposition of two oppositely travelling waves, at least two quantum numbers $\pm m$ are required for each component of angular momentum.

Since the possible numerical values of the angular momentum components are integers it appears reasonable that the modulus L, which classically is the maximum possible for any of the components, should obey a rule of the same form, *i.e.* $L = l\hbar$, l integer. This rule however, does not hold for quantum systems. Since

$$L^2 = L_x^2 + L_y^2 + L_z^2$$

the maximal value of L_z could reach L only if $L_x = L_y = 0$, which means that all of the components have well-defined values, which can only happen for $L = 0$ in a state of spherical symmetry. If the quantities L^2 and L_z are assumed to both have well-defined values the variables L_x and L_y can therefore not be dispersion free and for maximum L_z it always follows that

$$L^2 \leq L_z^2 + (\Delta L_x)^2 + (\Delta L_y)^2$$

Denoting by l the integer corresponding to the maximal value of L_z such that $L_z = l\hbar$, the inequality is of the same form as the rigorous quantum-mechanical result

$$L^2 = l(l+1)\hbar^2$$

If L^2 and L_z have well-defined values there will be $(2l+1)$ states corresponding to the angular momentum vector.

The heuristic description of Lévy-Leblond as outlined above provides a direct guide for the interpretation of angular momentum according to Bohm.

3.6.1 Bohmian Interpretation

Whereas linear momentum in the Bohmian interpretation relates to the translation of wavefronts ($\mathbf{p} = \nabla S$), angular momentum is described by their rotation. The components of angular momentum along a space direction (z-axis) are defined in a central field (*e.g.* hydrogenic atom) by

$$
\begin{align}
Y_{lm_\ell}(\theta, \phi) &= f_{lm_\ell}(\theta) \exp(im_l\phi) && (3.16) \\
L_z Y_{lm_\ell}(\theta, \phi) &= m_l\hbar Y_{lm_\ell}(\theta, \phi) && (3.17)
\end{align}
$$

where f_{lm_ℓ} are a set of real functions proportional to Legendre polynomials. A stationary state corresponding to energy E is given by

$$\Psi_{Elm_\ell}(r, \theta, \phi) = g_{Elm_\ell}(r) f_{lm_\ell}(\theta) \exp(i[m_l\phi - Et/\hbar]) \qquad (3.18)$$

where g_{Elm_ℓ} is real. [Compare $\Psi = R\exp(iS/\hbar)$]. The rotating phase func-

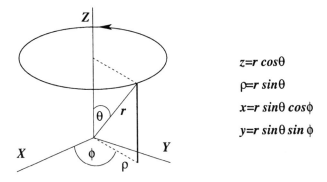

$z = r\,cos\theta$
$\rho = r\,sin\theta$
$x = r\,sin\theta\,cos\phi$
$y = r\,sin\theta\,sin\phi$

Figure 3.5: *Rotating wavefront and possible particle orbit for non-zero orbital angular momentum along z.*

tion, allowing for an arbitrary phase factor, hence is

$$S(r, \theta, \phi, t) = m_l\hbar\phi - Et + 2\pi n\hbar \, , \, n \in \mathbb{Z}$$

For given t and $m_\ell \neq 0$, the wavefronts $S = constant$, are planes parallel to, and ending on the z-axis with starting values $r_o\theta_o\phi_o$ as in figure 5. (z is a nodal line of Ψ when $m_\ell \neq 0$).

As t increases the planes rotate anticlockwise about the z-axis with angular velocity $\Omega = \dot{\phi} = E/m_\ell\hbar$, as shown in figure 6. As the phase changes

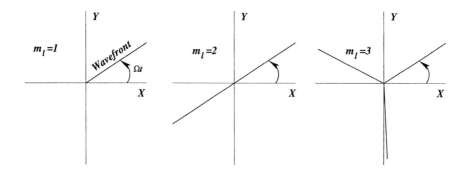

Figure 3.6: *Circulating wavefronts for various m_ℓ.*

during rotation single-valuedness of Ψ requires that

$$\oint dS = nh$$

and implies n nodal lines or n wavefronts within one cycle. In the present instance

$$m_l\hbar \int_0^{2\pi} d\phi = \pm m_\ell h \ , \ \text{whence} \ |m_\ell| = n \ .$$

The wavefronts for different values of $|m_\ell|$ are shown in figure 6. The particle (charge) orbits the z-axis along a circle of constant radius ($\rho_o = r_o \sin\theta_o$), which is not fixed as in the old Bohr theory. The wavefronts of states for which $m_\ell < 0$ rotate clockwise.

The number of rotating planes corresponds to $|m_l|$. For $m_l = 0$ the wave function (16) is real. In a spherically symmetrical environment there is no special direction in space and the wave function (18) acquires geometrical meaning only when an external magnetic field is switched on. Any of the

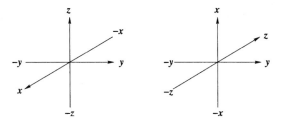

Figure 3.7: 90° *rotation of the coordinate axes about y.*

three degenerate *p*-states:

$$Y_1^1 = \left(\frac{3}{8\pi}\right)^{\frac{1}{2}} \sin\theta e^{i\phi} \propto \sin\theta\cos\phi + i\sin\theta\sin\phi = \frac{x+iy}{r}$$

$$Y_1^0 = \left(\frac{3}{4\pi}\right)^{\frac{1}{2}} \cos\theta \propto \frac{z}{r} \qquad\qquad (3.19)$$

$$Y_1^{-1} = \left(\frac{3}{8\pi}\right)^{\frac{1}{2}} \sin\theta e^{-i\phi} \propto \sin\theta\cos\phi - i\sin\theta\sin\phi = \frac{x-iy}{r}$$

may be designated to specify the *z*-direction, provided the corresponding eigenfunction is made real by an appropriate linear combination of eigenfunctions, such as

$$\sin\theta\left(e^{i\phi} + e^{-i\phi}\right) = \sin\theta\cos\phi = \frac{x}{r}$$

$$\sin\theta\left(e^{i\phi} - e^{-i\phi}\right) = \sin\theta\sin\phi = \frac{y}{r}$$

This procedure is equivalent to a rotation of the coordinate axes, as shown in figure 7. Rotation by 90° around the *y*-axis replaces *z* by *x*, *x* by −*z* and leaves *y* unchanged. When these substitutions are made in (19) eigenfunctions of L_x are obtained:

$$Y_1^1 \propto \frac{x+iy}{r} \rightarrow \frac{-z+iy}{r}$$

$$Y_1^0 \propto \frac{z}{r} \rightarrow \frac{x}{r}$$

$$Y_1^{-1} \propto \frac{x-iy}{r} \rightarrow \frac{-z-iy}{r}$$

These functions are still eigenfunctions of L^2, with $l = 1$ and the respective eigenvalues of L_x can be shown to be the same as the L_z of the original functions [49].

Once the special direction has been fixed, the two remaining eigenfunctions always constitute a complex pair with rotational symmetry in the xy-plane. A geometrical representation related to such functions is shown in figure 8. The familiar drawings of a set of three orthogonal p_x, p_y and p_z

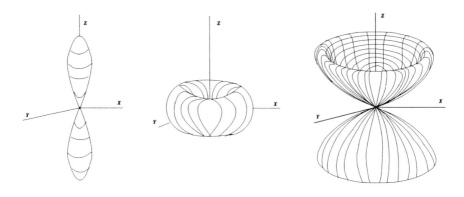

Figure 3.8: *Associated Legendre functions P_1^0, $P_1^{\pm 1}$, $P_2^{\pm 1}$ plotted as polar diagrams.*

orbitals to be found in many chemistry textbooks include the linear combinations that redefine the special z-direction along either x or y. The three real functions as a set, therefore has no physical meaning. Conventional hybridization schemes, invoked to rationalize the formation of multiple bonds, are likewise mathematical impossibilities. The generally accepted explanation of the rotational rigidity of double bonds then also has no physical basis.

Whenever an absolute direction can be defined, for instance in a molecule, the conservation of angular momentum as a manifestation of the rotational symmetry of space no longer holds. In practice this means that for any atom in an environment of less than spherical symmetry, an absolute direction exists, fixing the z-direction. In any degenerate set of states only one can be specified in real form; the others cannot be located more closely than to regions of rotational symmetry around the z-axis.

In practice this means that for an atom in any environment of symmetry lower than spherical, an absolute direction exists and fixes the direction of z. As a working rule it follows that only one of the three degenerate p-states (p_z) can be specified in real form. Other p-states in the system cannot be located more closely than to a plane perpendicular to z, with orbital angular momentum quantum numbers $m_l = \pm 1$. Non-zero m_l implies circulating charge and non-zero kinetic energy.

Since moving charges must by definition be less effective as chemical binding agents it is logical to expect residual orbital angular momentum to become quenched during the formation of chemical bonds. It is noted and demonstrated in table 2 that application of the same principle of quenching the angular momentum of motion leads to Hund's rule for the periodic variation of ground-state configuration of consecutive atoms with partially filled p, d or f levels. This quenching ensures that atoms remain spherically

Number of d-electrons	Orbital angular momentum, $m_l =$				
	2	1	0	−1	−2
1			↑		
2	↑				↑
3	↑		↑		↑
4	↑	↑		↑	↑
5	↑	↑	↑	↑	↑
6	↑	↑	↑↓	↑	↑
7	↑↓	↑	↑	↑	↑↓
8	↑↓	↑	↑↓	↑	↑↓
9	↑↓	↑↓	↑	↑↓	↑↓
10	↑↓	↑↓	↑↓	↑↓	↑↓

Table 3.2: *The orientation of electron spins as dictated by the requirement of quenched orbital angular momentum, illustrated for d-electrons, predicting Hund's rules.*

symmetrical, which is the most likely property at the root of Hund's rule.

A state with $l \neq 0$ but $m_l = 0$ may seem to represent the anomalous situation of a stationary electron with finite angular momentum. However, just like the potential energy has classical and quantum contributions, the total angular momentum also consists of two parts,

$$L^2 - \frac{\hbar^2 (\boldsymbol{x} \times \nabla R)^2}{R} = l(l+1)\hbar^2 \qquad (3.20)$$

$$L^2 = \frac{m_l^2 \hbar^2}{\sin \theta}$$

The second term on the left in (20) is explained as a quantum torque not associated with the motion and although $L^2 = 0$ for a stationary particle, the angular momentum may be non-zero, as in a p_z-state [35]. Although the electron has angular momentum in three dimensions the projection thereof on an axis, is zero. The quantum torque can therefore not be due to rotation about an axis and arises from spherical rotation about a point, to be described in chapter 4.7.

3.7   Chemical Significance

Bohmian formalism provides a much clearer link to chemical phenomena than the conventional interpretation. The remainder of this book concentrates on this topic.

Crystals, macromolecules, molecules, radicals, functional groups and atoms all represent various states of chemical aggregation that can be described by either a single or a product electronic wave function. Each wave function or factor in a product state has a characteristic quantum potential. This quantum potential stabilizes the system by balancing the classical potential. The fragmentation of an unstable entity is linked to the tendency of the wave function to factor into a product state. By way of illustration, a stable molecular crystal that represents a holistic unit is characterized by a single wave function and a fixed chemical potential. Within the crystal, the embedded molecules may formally be considered as smaller wholes, each with its own wave function and chemical potential. However, the holistic crystal is more than the sum of its molecular parts, since the product wave function ignores all intermolecular interactions responsible for the cohesion in the crystal. Molecular crystals with high vapour pressure have low quantum potentials. Such a crystal can be described by a product state, without substantial error. In this sense it is possible to distinguish between holistic and partially holistic chemical systems.

There are two important conclusions to be drawn from this insight, namely the additive nature of many molecular properties and the existence of non-local intramolecular interactions. Charge distribution and conformation of molecules are determined holistically by non-local interaction. All local features are consequences of the whole. However, molecules of any complexity are rarely the product of a one-step reaction that starts from the constituent atoms, but are more likely built up from intermediate fragments that retain some of their own molecular properties on incorporation into a bigger whole. This mechanism explains the large number of additivity rules that have been discovered empirically in molecular systems [51]. Diverse additive properties include molar volume, parachor, refractivity, polarizability, heat of formation, magnetic rotation (Faraday effect) and diamagnetic susceptibility. A molecule whose conformation and properties are functions of its chemical history is not holistic, but partially holistic [2], which means that its wave function is a product function, albeit with a limited number of factors (fragments)

$$\Psi = \prod_{i=1}^{n} \phi_i \quad \text{where} \quad 1 < n < a \text{ , the number of atoms in the molecule.}$$

The non-local holistic interaction is a function of fragment size (*i.e.* the quantum potential). When a fragment loses its integrity under the influence of holistic interaction, intramolecular rearrangement occurs. The rearranged structure is an emergent property.

All thermodynamic and electronic properties of molecules are closely linked to the quantum potential. Many of these, for instance electronegativity, only known from empirical relationships before, can now be demonstrated to be of fundamental theoretical importance. The close similarity between chemical potential of a system and the quantum potential of component molecules establishes a direct link between quantum mechanics and thermodynamics, without statistical considerations. This relationship has direct bearing on the nature, mechanism and kinetics of chemical bond formation, including sterically improbable intramolecular rearrangements.

The second major link between Bohmian mechanics and modern chemistry concerns the understanding of molecular shape. A quantum-mechanical molecule is a holistic unit of nuclei and electrons [7] characterized by a molecular Hamiltonian. The Hamiltonian specifies all situations and permutations consistent with a given partitioning of n particles into α nuclei and i electrons (T7.27) and only a minor fraction of these corresponds to classical stereochemical forms. The molecular wave function in $3n$-dimensional configuration space has a rich enough mathematical structure to describe all conformations of all possible isomers, as well as fragmentation products consistent with the molecular formula. In fact, the molecular wave function describes all possible distributions of the n particles and a specific solution can only be the result of an equally specific set of boundary conditions, which depend not only on the physical environment of the system, but also on the history of the particles as a set. It is well known that, without exception, the constituent fragments of a molecule may exist, without change, under exactly the same conditions as the molecule itself. The two systems are described by the same wave function and only have different histories.

The stubborn believe that each molecular wave function relates to a

unique structure that may be extracted by energy-minimization methods
of conventional quantum chemistry, is baseless. It is only in Bohm formalism
that the definition of molecular shape, becomes feasible. The unifying concept in this instance is the orbital angular momentum of molecular fragments.
Minimization of residual torque, rather than electronic energy is found to be
responsible for the generation of molecular shape. It makes more sense to
find that three-dimensional structure relates to a vector, rather than a scalar
quantity. The important stereochemical effects of photochemical reactions
emerges as a consequence of angular momentum transfer between photons
and molecular matter.

The elusive relationship between molecular chirality and optical activity,
as well as the Faraday effect, likewise reduce to the minimization of orbital
angular momentum as a function of molecular symmetry.

The most revolutionary consequence of the Bohmian interpretation of
orbital angular momentum is perhaps the natural interpretation of non-steric
molecular barriers to rotation and the nature of multiple bonds. It completely
destroys the notion of π-bonding and explains the absence of a barrier to
rotation in bonds of odd integer order.

Chapter 4

Structure of the Electron

4.1   Introduction

Many of the concepts common to chemistry and physics are used so differently in the two disciplines that the common ground is often hard to recognise. One such a topic is thermodynamics, which both disciplines borrowed from engineering and still have not managed to incorporate into their distinctive paradigms. Another is the electron. Its appearance was keenly anticipated in both camps before its actual discovery, albeit for different reasons. Chemists were looking for the unit of electricity, named electron [52], that links the molar constants of Faraday and Avogadro. In the physical theory of matter the isolated electric charge was postulated as the key to explain inertial mass.

The expectations of electrochemists were met by discovery of the electron which soon led on to other exciting innovations to explain chemical bonding as electronic interaction. In physics however, the story was different, since the electron failed to meet the simultaneous dictates of electromagnetism, special relativity and mechanics, even before the complicating results of the photoelectric effect and electron diffraction became known. The situation has not changed much following the development of quantum theory that came into being primarily to provide an improved understanding of the electron.

Despite a lot of posturing the electron of chemistry is still the electron of Lewis [53], untouched by quantum electrodynamics (QED). The lip service paid to wave mechanics and electron spin, even in elementary chemistry textbooks, does not alter the fact that the curly arrow of chemistry signifies no more than redistribution of negative charge. By holding out the prospect of an intelligible structure of the electron, quantum mechanics created the expectation that chemistry could be reduced to a subset of physics, explaining all chemical interactions as quantum effects. The result of this unfulfilled promise is that chemistry is without a theory today. It is true that physics and chemistry both study the relationship between matter and energy but, while physics is concerned with matter that moves, chemistry examines how matter changes from one form into another. The subject matter may be the same but the emphasis is different.

Because of the inadequacies of QED a fundamental theory of electrode processes is still lacking. The working theories are exclusively phenomenological and formulated entirely in terms of ionic distributions in the vicinity of electrode interfaces. An early, incomplete attempt [54] to develop a quantum mechanical theory of electrolysis based on electron tunnelling, is still invoked and extensively misunderstood as the basis of charge-transfer. It is clear from too many superficial statements about the nature of electrons that the symbol e is considered sufficient to summarize their important function. The size, spin and mass of the electron never feature in the dynamics of electrochemistry.

Theoretical chemistry must search for realistic models and alternative interpretations of basic theory that accord with the time-honoured empirical concepts of chemistry. Above all chemistry needs a theory of the electron, the one object at the heart of all chemical change.

4.2 Historical Development

For a hundred years since its discovery the electron has provided a topic for intense study and speculation. It has a fixed charge, mass and spin: each of these is of fundamental importance. The discovery has spawned an enormous electronic industry and the science of chemistry is to a large extent a study of electron distributions and transfer. There is still no consistent theory of the electron. Particle physicists - the ultimate experts - simply describe it as a stuctureless point particle. In the context of atomic physics and chemistry this description is meaningless in view of the demonstrable structure of atoms, molecules and crystals, largely derived in terms of the classical Thomson model of scattering at an electron. Following Jackson [55] the scatter of electromagnetic radiation ($e.g.$ X-rays) from an electron defines a differential scattering cross section

$$\frac{d\sigma}{d\Omega} = \frac{\text{energy radiated/unit time/unit solid angle}}{\text{incident energy flux in energy/unit area/unit time}}$$

For unpolarized incident radiation, using electrostatic units,

$$\frac{d\sigma}{d\Omega} = \left(\frac{e^2}{mc^2}\right)^2 \cdot \frac{1}{2}(1 + \cos^2\theta)$$

The total scattering cross section, called the Thomson cross section,

$$\sigma_T = \frac{8\pi}{3} \left(\frac{e^2}{mc^2} \right)^2$$

The unit of length, $e^2/mc^2 = 2.82 \times 10^{-13}$ cm, is called the classical electron radius, because a classical distribution of charge totalling the electronic charge must have a radius of this order for its electrostatic self-energy to be equal to the electronic mass energy. This estimate of the size of an electron is crude, but still the only one available.

The earliest classical theories were based on an electron considered a spherical atom of electricity and of mass, entirely electromagnetic in origin. According to a relativistic (Lorentz) version of this theory one assumes a flexible spherical charge distribution as the concept of rigidity is not relativistically invariant. This explanation was only partially successful because the various parts of such a charged sphere should dissipate in space due to Coulombic repulsion. The solution (Poincaré,) [56] to this instability is that non-electromagnetic cohesive forces within the electron hold it together. The observed electronic mass

$$m_o = m_c + m_e \tag{4.1}$$

therefore consists of cohesive and electromagnetic parts that can not be observed independently. The idea of electromagnetic origin of mass was finally abandoned after discovery of the neutron.

The new mass assumption (1) removed the need of an extended charge distribution formerly required to generate mass and opened the possibility of a classical point-charge description of the electron. This model was examined by Dirac [57] who managed to overcome the problem of infinite Coulombic energy by symmetrizing the electronic field on combining (temporally) advanced and retarded field components.[1] The results agreed with the earlier relativistically Lorentz invariant description, but in the physical interpretation the finite size of the electron reappeared in a new sense. Whereas the point-charge model requires an acausal pre-acceleration this condition is avoided if the electron has three-dimensional size with an interior through which signals can be transmitted at a speed greater than that of light.

The relativistic Lorentz-Dirac description of the electron is generally accepted to be the ultimate classical description, albeit without Dirac's physical

---

[1]This "renormalization" procedure with its clear physical basis has not been generally accepted because of the implied "nominally acausal effects" [58], but it provided a basis of the Wheeler-Feynman absorber theory and the discussion that follows here.

interpretation. It is argued that only a quantum theory can reveal the internal structure of matter [55], but a theory of elementary particles generally accepted as adequate has not yet emerged from quantum electrodynamics. This situation is not entirely surprising in view of the fact that QED is an explicit relativistic theory of assumed point sources of the electromagnetic field. In the words of Dirac [59] QED succeeds "in setting up rules for handling the infinities and subtracting them away, so as to leave finite residues which can be compared with experiment, but the resulting theory is an ugly and incomplete one, and cannot be considered as a satisfactory solution of the problem of the electron."

4.3   The Quantum View

It is reasonable to demand that a successful quantum-mechanical model of the electron should contain and improve on well established features of the classical electron. These include a reasonable cross section for X-ray scattering (Thomson), a non-rigid structure (Lorentz), a well defined mass, not of electromagnetic origin, associated with a characteristic charge (Poincaré), and correct invariances, including a non-local interior region (Dirac). The first objective of quantum theory, in its various guises has been to provide a detailed description of the electron in all possible situations. The only exception is quantum theory according to Heisenberg's formulation in which minimal physical content is associated with the mathematical variables of the theory; all other formulations are wave-mechanical and plagued by the ever present wave-particle dilemma. The no-nonsense consensus among hard-nosed theoreticians with no other ambition than to perform the ultimate calculation is that wave functions are mathematical constructs in configuration space, without physical meaning in real space and in terms of which the physics of point-particles can be analyzed. This view is equivalent to the Heisenberg approach and is accepted by most quantum theorists in the tradition of the Copenhagen school. According to their dogma there is no scope for philosophical speculation about the meaning of wave functions or the internal structure of electrons and photons.

As a cultural activity this harsh discipline of quantum physics has been a failure. The layman continues to ask about the meaning of things and an occasional heretic, such as David Bohm [60], keeps the idea of an intelligible quantum reality alive. The quest for an electronic structure must be conducted in this spirit: one can scarcely expect a discipline based on the electron as a zero-dimensional point to reveal its structure. However, despite Dirac's initial neglect of dimensional properties, he found a finite classical electron. As the electron is known to have mass, charge and spin, it is axiomatic to have physical extent, and quantum theory is expected to reveal its structure.

In the most widely used form of wave mechanics the Schrödinger equation describes an electron in terms of non-relativistic wave functions. Uninformed commentators regularly criticise this formulation for its inability to account for electron spin and immediately turn to relativistic equations. This reaction seems to be vindicated by the way in which relativistic wave equations, in addition, produce the negative-energy solutions characteristic of anti-matter. What is not generally appreciated is that spin operators are unique not to relativistic Hamiltonians but to Hamiltonians of correct symmetry, be it Lorentzian or Galilean. The linearized Galilean invariant non-relativistic Schrödinger Hamiltonian contains the same spin operators as the Dirac Hamiltonian. Furthermore, the commonly used Schrödinger operator

$$i\hbar\frac{\partial}{\partial t} - \frac{\hbar^2\nabla^2}{2m}$$

is equally valid in its complex conjugate form and hence represents two separate solutions for matter and anti-matter. This result is obvious [61] because Dirac's equation goes into Schrödinger's equation in the limit of infinite luminal velocity, *i.e.* $c \to \infty$. In this sense it is superior to the Dirac equation that mixes positive and negative energy states in one spectrum, separated by a gap $2mc^2$. The Schrödinger equations define an unbridgeable gap and generate separated energy spectra, as idealized but never achieved by Dirac, who stated [62] that "the true relativity wave equation should be such that its solutions split into two non-combining sets, referring to the charge $-e$ and the charge e".

4.3.1 Hole theory

The problem remains unresolved to this day. It makes no sense to assign the negative-energy solutions to positive-charge particles unless there is an energy barrier preventing electrons from cascading down the energy ladder to negative infinity energy. To prevent this, Dirac postulated the infinite negative-energy sea of electrons that occupy all of these states to exclusion, according to the Pauli mechanism. Any accidental holes could be ascribed to positively charged particles. Dirac's first proposal of this kind identified holes with protons, later changed to anti-electrons. The problem with this proposal was to defend the notion of infinite mass and charge in any finite volume. Despite this difficulty the implied concept of pair creation and annihilation gained such importance that the excesses of the original proposal were simply ignored in the subsequent development of quantum field theory. This approach reflects the true spirit of the Copenhagen school: to avoid the embarrassment of the physically infinite sea it was replaced by a field of creation and annihilation operators with calculated energies and charges trimmed by infinite mathematical correction factors. According to Saunders [63] this new 'correct' particle interpretation of the field no longer demands a physical interpretation of its normal-ordering process. This approach is to be considered a purely mathematical technique which stands no need of justification.

4.3.2 Zitterbewegung

An alternative interpretation of the positive and negative eigenfunctions of the Dirac Hamiltonian [65] was sought by Schrödinger [66] by introducing the potential-energy term in a form that leads to the mixing of positive and negative functions. The result is a negligibly small component, superimposed on the normal eigenfunctions.

The small component was first identified by Schrödinder [67] when solving the Heisenberg equation of motion (T5.1.7)

$$\frac{h}{2\pi i}\frac{\mathrm{d}A}{\mathrm{d}t} = HA - AH$$

in terms of the Dirac Hamiltonian (T5.3.1). The solution for any coordinate,

$$\left(\frac{\mathrm{d}x_k}{\mathrm{d}t}\right)^2 = c^2 \cdot I$$

where I is a unit matrix, indicated that the electron moved at light velocity. To explain this result it was further shown that the calculated velocity could be decomposed into a linear and an oscillatory factor respectively corresponding to the group velocity v_g of a De Broglie wave packet of momentum p_k, and a high-frequency oscillation at the Compton wavelength λ_C. Compare (T3.6.4). From the product of group velocity and phase velocity v_ϕ of dispersive waves

$$v_g v_\phi = c^2$$

it follows that $v_\phi > c$, and this is the superluminal component associated with Zitterbewegung.

Although the behaviour of the Dirac electron appears not to be affected by the small component, it means that despite its moderate velocity the electron acquires velocity components of order c. It was concluded that the electron is in non-linear motion, with a high-frequency periodic trembling motion (Zitterbewegung)[2] of amplitude $h/4\pi mc \simeq 10^{-13}$ m (the Compton wavelength) superimposed on the perceived smooth linear progression. This amplitude only represents the mean displacement of the centre of gravity and not the size of the electronic charge cloud. Once again the finite size of the electron manifested itself despite assumptions to the contrary. It is significant to note that the Zitterbewegung of a non-relativistic Schrödinger electron $(c \to \infty)$ could have velocities in excess of c, like the classical Dirac electron.

It was Schrödinger's intention to associate Zitterbewegung with electron spin, but such an assumption would serve simply to clarify one mystery in terms of another. Instead, one could try first to understand the nature of Zitterbewegung within the region identified as the electron. The only substance available to support the periodic motion is space itself and there seems to be two possibilities: either space consists of continuous stuff or of compacted particles. Winterberg [68] explored the latter possibility. Wave motion in a continuous aether is probably easier to visualize and needs fewer assumptions. The only postulate is that ponderable matter and its properties represent special configurations of space. Hence flat Euclidean space (-time) in dimensions of any number, is featureless and empty.

Particle model

According to Winterberg's model the vacuum (or Planck aether) consists of a lattice made up of an equal number of elementary positive and negative Planck masses, $(m_p = \sqrt{\hbar c/G})$, the size of a Planck length r_p, $(m_p r_p c = \hbar)$, and with zero gravitational effect. In the classical limit, the Planck aether can be described entirely by waves with uncertainty relations of classical mechanics (T3.6.3). Quantization amounts to the establishment of standing wave patterns with all particles and physical objects represented by wave packets of the Planck aether. The quantum conditions $E = \hbar\omega$ and $p = \hbar k$ lead to the uncertainty relations $\Delta p \Delta q \geq \hbar$, $\Delta E \Delta t \geq \hbar$, resulting in a zero-point energy. Quantum fluctuations of the positive mass aether are equal and opposite to fluctuations of the negative mass aether, thereby cancelling each other. From this point onwards the theory is developed entirely in terms of the wave model and the Planck masses never feature again. In fact, the Planck aether is viewed as composed of two superfluids, one with a positive

---

[2]Quantum field theory is claimed to have been developed in an effort to explain Zitterbewegung. The successes of the programme cannot be denied, but a lacking achievement is to establish the structure of the electron.

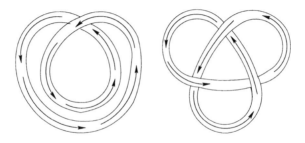

Figure 4.1: *Schematic wave forms representing electrons and protons.*

sically stable particles would be formed from only a few precise frequencies of radiation. In a later communication Jennison [75] considered possible structures in more detail; he suggested that cavities that trapped electromagnetic radiation with a single node would best reproduce the observed properties of electrons and protons, tentatively represented in one dimension by rotating standing waves of the type shown in figure 1. In terms of the present discussion it would be premature to introduce electromagnetic waves, but, the principle of inertia associated with a locally trapped standing wave can safely be grafted on preceding arguments without specifying the nature of radiation.

An example of an electron in a phase-locked cavity has been encountered in the study of compressed atoms [76, 24]. Isotropic compression of an atom, simulated by imposing a finite boundary condition on the electronic wave function, *i.e.* $\lim_{r \to r_0} \psi_e = 0, r_0 < \infty$, raises electronic energy levels, until an electron is decoupled from the core at a characteristic atomic ionization radius. This electron then exists in a field-free cavity with a spherical Bessel wave function. In the ground state

$$\psi = \frac{A \sin kr}{kr}$$

with $k = \sqrt{\frac{2mE}{\hbar^2}}$.

The important conclusion is that the electron has no fixed size; in three dimensions it fills any available cavity as allowed by the environment, in the form of a standing spherical wave. In this sense there is no such thing as a completely free electron. When stated in the following that the state of an electron depends on all other matter in the universe, this may well be the case, but in practice it probably refers to no more than a local neighbourhood.

4.5   Wave Model of the Electron

The ripple that occurs in undistorted space should be isotropic and of a type described with a general wave equation

$$\nabla^2 U = \frac{1}{v^2}\frac{\partial^2 U}{\partial t^2} \tag{4.2}$$

If U is a function of the radius vector and t alone, ∇^2 reduces [77] to

$$\frac{\partial^2}{\partial r^2} + \frac{2}{r}\frac{\partial}{\partial t^2}$$

and the equation reads

$$\frac{v^2}{r}\frac{\partial^2(rU)}{\partial r^2} - \frac{\partial^2 U}{\partial t^2} = 0$$

The substitution $\xi = r + vt, rU = P$ converts it into

$$v^2\frac{d^2 P}{d\xi^2} - v^2\frac{d^2 P}{d\xi^2} = 0 \quad ;$$

hence $P = f_1(r+vt)$. A similar result would have been achieved by choosing $\eta = r - vt$ in place of ξ. Hence

$$P = f_1(r + vt) + f_2(r - vt)$$

or

$$U = \frac{1}{r}[f_1(r + vt) + f_2(r - vt)]$$

This solution represents two spherical waves (T 5.5.6), one travelling toward the origin, the other from the origin. The factor $1/r$, without which U would not be a solution of (2) and therefore not a wave, accounts for attenuation of a spherical wave as it moves from its source. By suitable choice of f_1 and f_2 diverse wave complexes can be formed, of which standing waves, defined with a condition $U(r,t) = F(r) \cdot G(t)$ in which F and G represent new functions, are perhaps the simplest. Allowing for time-inversion symmetry a more general solution of (2) is

$$U = \frac{U_o}{r}\exp\left[\pm i(\mathbf{k} \cdot \mathbf{r})\right]\exp\left[\pm 2\pi i\nu t\right]\exp(i\phi) \tag{4.3}$$

in which $\mathbf{k}$ is the wave vector ($k = 2\pi/\lambda$) of the wave, pointing in the direction of propagation, ϕ is an arbitrary phase and ν is the frequency, ($v = \lambda/\tau =$

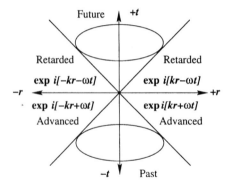

Figure 4.2: *Minkowski diagram of four solutions of wave equation (3) for an emitter at the origin.*

$\lambda\nu$). Cramer [78] plotted these solutions on a Minkowski diagram, adapted here as figure 2. Returning to the notion of a particle as a standing wave in a phase-locked cavity, it is noted that a standing wave constructed from diverging and converging retarded waves corresponds closely to the construct of Jennison and Drinkwater [73]. *i.e.* $\varepsilon(r, t) = u(r)\psi(t)$, in which

$$\Phi = A\frac{\varepsilon(r, t)}{r} = \frac{A}{r}e^{-i(\omega t + kr)} - \frac{A}{r}e^{-i(\omega t - kr)}$$

$$= \frac{A}{r}e^{-i\omega t}\left[e^{-ikr} - e^{ikr}\right]$$

$$= \frac{2iA}{r}e^{-i\omega t}\sin kr$$

$$\Phi = \frac{\Phi_o \sin k_o r}{k_o r}e^{-i\omega_o t} \tag{4.4}$$

in which $\omega_o = k_o c$, the subscript defines a state of rest. It is next assumed that incoming and outgoing waves have frequencies $\omega_1 = k_1 c$ and $\omega_2 = k_2 c$ respectively, to yield

$$\varepsilon = e^{i(k_1 r - \omega_1 t)} - e^{-i(k_2 r + \omega_2 t)}$$

$$= 2\sin\left\{\frac{k_1 + k_2}{2}r - \frac{\omega_1 - \omega_2}{2}t\right\}$$

$$\times\left[-\sin\left\{\frac{k_1 - k_2}{2}r - \frac{\omega_1 + \omega_2}{2}t\right\} + i\cos\left\{\frac{k_1 - k_2}{2}r - \frac{\omega_1 + \omega_2}{2}t\right\}\right]$$

Following Elbaz [79] the following substitutions are made:

$$\omega = \frac{\omega_1 + \omega_2}{2} = \frac{k_1 + k_2}{2}c = kc$$

$$\beta = \frac{\omega_1 - \omega_2}{\omega_1 + \omega_2} \quad , \quad \omega_o^2 = \omega_1 \omega_2$$

whereby

$$\varepsilon(r,t) = -2i \sin\{kr - \beta\omega t\} \exp\{-i(\omega t - \beta k r)\}$$

As $kr - \beta\omega t = k_o r'$, $\omega t - \beta k r = \omega_o t'$, it follows that

$$\varepsilon(r,t) = 2u(r,t)\psi(r,t)$$

now defines a standing wave in a moving cavity

$$\Phi = \frac{\Phi_o}{r} \sin(k_o r') e^{-i\omega_o t'}$$

In this,

$$r' = \frac{r - (\beta/k)\omega t}{k_o/k} = \frac{r - vt}{k_o/k} \quad , \quad \text{writing } \beta = v/c.$$

$$\frac{k_o}{k} = \frac{2\sqrt{\omega_1 \omega_2}}{\omega_1 + \omega_2} = \sqrt{1 - \beta^2}.$$

This result is consistent with the Lorentz transformations

$$r' = \frac{r - vt}{\sqrt{1 - \beta^2}}$$

$$t' = \frac{1}{\sqrt{1 - \beta^2}} \left(t - \frac{vr}{c^2} \right)$$

Waves in the moving cavity therefore lock together in a relativistically invariant way.

In order to reconcile this relativistic invariance with De Broglie's wavelength hypothesis, for waves at rest and in a moving frame,

$$\psi = \exp(2\pi i \nu_o t_o) = \exp[2\pi i(\nu t - r/\lambda_B)]$$

Elbaz [80, 81, 82, 83] indicated that the assumption is incomplete and that an amplitude wave

$$u = \exp(2\pi i r_o/\lambda_o)$$

served to give physical meaning to the relation $E_o = m_o c^2$. Thus, the standing wave associated with a massive particle, $\varepsilon(r,t) = u(r)\psi(t)$ is expected to include both the De Broglie wave and an amplitude wave of length,

$$\lambda_o = c/\nu_o = hc/h\nu_o = hc/m_o c^2 = h/m_o c,$$

which is the Compton wavelength, *i.e.*

$$u = \frac{B}{r} \exp[2\pi i(r/\lambda_C - Nt)]$$

in which the frequency $N = mvc/h$. The plane-wave solutions of function u have a wavelength equal to the Compton wavelength $\lambda_C = h/mc$, a phase velocity equal to the particle velocity v, and a group velocity c^2/v, whereas the quantum-mechanical ψ-function has a wavelength equal to the de Broglie wavelength $\lambda_B = h/mv$, a phase velocity c^2/v and a group velocity v. Hence the material particle at rest is associated with a standing wave obtained by the superposition of two running waves propagated in opposite directions. When the two frequencies are identical the resulting standing wave corresponds to the particle at rest. When the frequencies differ from each other, a standing wave moving with a velocity

$$v = c\frac{\omega_1 - \omega_2}{\omega_1 + \omega_2}$$

is obtained. Wolff showed independently [84] that the superposition of two spherical waves, respectively moving from and toward the origin, produced the result of equation (4). Choosing $\nu = \omega/2\pi$ equal to mc^2/h, $k = 1/\lambda_B$, one obtains a standing wave with the same properties as Elbaz's construction.

A radial diagram of the standing wave proposed [84] to represent an electron is shown in figure 3. Each wave packet represents a De Broglie wave and the elementary waves correspond to Zitterbewegung at the Compton frequency mc^2/h. The amplitude represents the electric potential of the electron. At the origin it corresponds to Φ_o, the limiting value as $r \to 0$. The charge-force decreases as $\Phi^*\Phi = 1/r^2$, except at the origin where it is finite and the Coulomb potential gives the wrong value. It is not too difficult to understand that the deviation from Coulomb's law, near the origin happens at the point where the quantum potential balances the classical potential, as shown in figure 3.4. It has been suggested [84] that this modification of Coulomb's law is equivalent to the renormalization of quantum field theory that eliminates unwanted infinities. In three dimensions the electron is represented by a modulated spherical wave, concentrated around the origin.

In terms of the present argument an elementary unit of distorted space (a particle) disturbs the universal equilibrium and becomes a source of secondary outward waves that interfere in space with all other waves produced by comparable sources. When the magical universal equilibrium becomes restored the resultant wave derived from all other matter in space-time con-

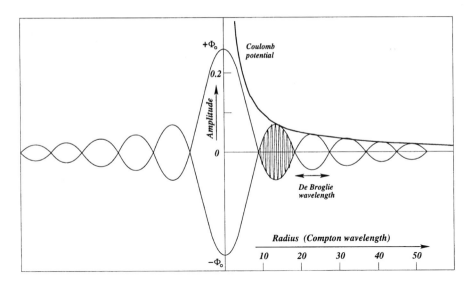

Figure 4.3: *Standing wave to represent an electron, obtained by the superposition of two spherical waves moving from and towards the origin respectively, after Wolff [84]. The envelope of the wave amplitude matches the Coulomb potential (1/r), except at the origin where it is finite.*

stitutes an in-wave at the position of that particle[3]. This condition is exactly that already considered of a pair of in/out waves interfering to yield a standing wave (4), which represents an electron with maximum amplitude at the origin. The same spherical Bessel function of an electron released by a compressed atom is part of the solution, which in addition has a more rapidly oscillating exponential component, readily identified with Zitterbewegung.

The mysterious phase velocity of the de Broglie wave and the group velocity of the amplitude wave, $c^2/v > c$, refer to the, by now familiar superluminal motion in the interior of the electron. As many authors noted and Molski(1998) recently reviewed [86] an attractive mechanism for construction of dispersion-free wave packets is provided in terms of a free bradyon[4] and a free tachyon that trap each other in a relativistically invariant way. It is demonstrated in particular how an electromagnetic spherical cavity may be

---

[3]It has been interpreted by Wolff [85] as a manifestation of Mach's principle of cosmic influence deriving inertial mass from the effect of distant giant masses.

[4]Object with velocity $< c$. A tachyon has $v > c$.

employed as a wave-corpuscular model of extended massive particles endowed
with spherical geometry, writing, as in (4)

$$\Psi_n(t,r) = E_o \frac{\sin \kappa r}{\kappa r} \exp[i\kappa t]$$

for integer n, in a cavity of radius r, $\kappa = \frac{n\pi}{a}$. A non-zero rest mass

$$m_o = \frac{\hbar \kappa}{c}$$

is attributed to the imprisoned fields in accordance with the model of Jennison and Drinkwater [73]. The trapped radiation undergoes conversion into a
time-like (B) and a space-like (D) wave that lock to form a solitary photon-
like wave Ψ_n. Such imprisoned inner fields form a nondispersive wave packet
that has no spread with time ($\kappa = m_o c/\hbar$, is constant) and travels at a group
velocity equal to that of the moving cavity. The diameter of the cavity

$$d = 2a = \frac{hn}{m_o c} = n\lambda_o$$

with λ_o the Compton wavelength characterising the cavity at rest. This
result is the precise analogue of Schrödinger's Zitterbewegung. The mass
associated with the cavity varies as

$$m = \frac{m_o}{\sqrt{1 - \frac{v^2}{c^2}}}$$

which indicates that radiation trapped in a spherical cavity behaves as ordinary ponderable matter.

The bradyon-tachyon coupling was first developed by Corben [87], in
terms of trapped masses rather than standing waves. The two objects trap
each other in a relativistically invariant way yielding a compound particle of
rest mass $M_o = \sqrt{m_o^2 - \mu_o^2}$, in which m_o and μ_o are respectively the masses
of a bradyon and a tachyon.

4.6   Electronic Charge

The inertial mass of the electron is considered adequately described in terms
of its wave structure, which may reasonably be expected also to elucidate the
properties of charge and spin. The most obvious feature that might relate
to charge or spin is the Zitterbewegung, or the Compton (tachyonic) part of
the trapped standing wave. This trembling by itself has no obvious physical

basis apart from being a property of an elementary distortion (vortex) in space-time.

Klein made an intriguing suggestion about the nature of electric charge [88] on the basis of a five-dimensional unified-field theory.

4.6.1 Kaluza-Klein Theory

The theory of Kaluza and Klein [89, 90] is based on an observation that of two macroscopic forces of Nature only gravitation can be ascribed to geometric features of four-dimensional space-time. In order to incorporate another interaction the logical development would be to consider an additional dimension and to examine if extra degrees of freedom provided by 15 covariant components of the five-dimensional symmetric tensor needed to specify the line element

$$d\sigma = \sqrt{\sum \gamma_{ik} dx^i dx^k} \quad , \quad i, k = 0, 4,$$

can be used to characterize the electromagnetic field. Four coordinates, $x^i (i \neq 4)$ say, must still characterize regular space-time and γ_{ik} must be independent of x^4. To explain why the fifth dimension (which may be called the *klein* dimension) is not observed, space was proposed to have cylindrical symmetry with respect to x^4,

$$\frac{\partial \gamma_{ik}}{\partial x^4} = 0.$$

On this basis it was possible to obtain equations which, to first approximation, agreed with known relativistic equations of a gravitational field and generalized Maxwell equations of electromagnetism. Charged particles in the electromagnetic field follow geodesic paths of five-dimensional space. Only fourteen coefficients serve to represent ten gravitational and four electromagnetic potentials. The meaning of γ_{44} is undefined. In a demonstration to derive Schrödinger's equation from a general five-dimensional wave equation it was shown that introduction of the quantum condition was closely connected with an assumed periodicity in x^4. The origin of Planck's quantum was conjectured to reside in this periodicity in the fifth dimension.

In this theory the equations of motion of an electrified particle are geodesics referred to a line element

$$d\sigma = \sqrt{(dx^4 + \beta\phi_i dx^i)^2 + g_{ik} dx^i dx^k}$$

in which the $x^{i \neq 4}$ are coordinates of ordinary space-time with line element $g_{ik} dx^i dx^k$, x^4 is a fifth coordinate, ϕ_i are four covariant components of an electromagnetic vector potential, and $\beta^2 = 2\kappa$, κ being Einstein's gravitational constant. With $d\tau$ as the differential of proper time belonging to a particle of mass m and charge q, the Lagrangian $\mathcal{L}$ for geodesics representing particle motion in five dimensions has the form

$$\mathcal{L} = \frac{1}{2}m \left(\frac{d\sigma}{d\tau} \right)^2$$

with momenta defined conventionally,

$$p_i = \frac{\partial \mathcal{L}}{\partial(\partial x^i / \partial \tau)} \quad , \quad (i = 0, 3),$$

but with p_4 constant along the geodesic, as x^4 is absent from $\mathcal{L}$. To arrive at the correct equations of motion one must set

$$p_4 = \frac{q}{\beta c}.$$

If q is an integral multiple of some elementary electronic charge e, $p_4 = Ne/\beta c$. If five-dimensional space is assumed to be closed in the direction of x^4 with a period l, p_4 is quantized according to

$$p_4 = \frac{Nh}{l} = \frac{Nh}{2\pi R}$$

N becoming a quantum number, which may be a positive or negative integer according to the sense of motion in the direction of the klein dimension; R is the radius of this compacted dimension. The relationship

$$e = \frac{\hbar c \beta}{R}$$

is considered to define an electronic charge as arising from circulation along the klein coordinate, or alternatively to determine the radius of the compact space[5]

$$R = \frac{\hbar(16\pi G)^{\frac{1}{2}}}{ec} = \frac{r_p}{\sqrt{\alpha}}$$

in which $\alpha = e^2/\hbar c$ is the fine structure constant and the Planck length is $r_p = 1.6 \times 10^{-33}$ cm. Displacement along x^4 implies rotation; hence the electronic

---

[5]An intriguing correspondence exists between the klein radius and vortices that feature in theories of Winterberg and Meno.

charge can also be interpreted [91] as angular momentum $p^4 R$ conjugated to an angular variable θ of period 2π. An operator on displacement of the klein coordinate might consequently be associated with electric charge $q \rightarrow -i\hbar\partial_4$ with positive and negative eigenvalues corresponding to the charge $\pm e$. Hence inversion of the x^4 variable is equivalent to charge conjugation [92], $\Psi(x^i) \rightleftharpoons \Psi(x^i)^\star$. As summarized by Molski [91] the Kaluza-Klein approach unifies gravity and electromagnetism; only one force acting on a charged particle is assumed. It manifests itself as gravity when a particle travels as a geodesic in the 4-space and as electromagnetism when it moves in a helical geodesic around the hyperthread in the compacted fifth dimension.

A five-dimensional wave equation based on Kaluza-Klein theory in the absence of gravitation has the form

$$\left[\Box^5 + (M_o c/\hbar)^2\right]\Psi(x^i) = 0 \qquad (4.5)$$

$$\Box^5 = \partial_i \partial^i = \partial_0^2 - \partial_1^2 - \partial_2^2 - \partial_3^2 - \partial_4^2$$

By dimensional reduction and variational methods the field equation in 4-space and the absence of fields

$$\left[\partial_\mu \partial^\mu + (m_o c/\hbar)^2\right]\Psi(x^\mu)_n = 0 \qquad (4.6)$$

Molski showed [91] that (5) and (6) can be viewed as a five-dimensional version of Corben's tachyonic theory. In particular $\Psi(x^\mu)$ may be interpreted as a B-wave associated with a bradyon of mass m_o moving in four-space M^4 wheras $\Psi'(x^4)$ is a purely space-like D-wave associated with a transcendent tachyon of mass μ_o and infinite speed moving in compact space S^1 about the bradyonic constituent and satisfying a wave equation

$$\left[\partial_4^2 - (\mu_o c\hbar)^2\right]\Psi(x^4) = 0$$

These two free objects trap each other in a relativistically invariant way, yielding a bradyon-tachyon compound of mass[6]

$$M_o = \sqrt{m_o^2 - \mu_o^2}$$

described with a wave function

$$\Psi(x^i) = \psi(x^\mu)\psi'(x^4).$$

---

[6]With this observation a conjecture of Poincaré embodied in (1) is finally substantiated.

4.7   Geometrical Model of the Electron

The assumption that matter is merely a distortion of the featureless fabric
of space-time, is in accord with Descartes, Kelvin, Clifford, Weyl, Einstein,
Eddington, Schrödinger and others. Despite many speculations, a consistent
reconstruction of the physical world as a special configuration of space-time
has never been achieved, but conspicuous progress, largely ignored by the
scientific community, has been made. The first problem [93] is to characterize
a single particle that preserves its shape relative to other regions of space
during motion. This problem can be considered solved by Battey-Pratt and
Racey [94] who developed a geometric model of fundamental particles in
terms of *spinors*. The spinor concept developed from the analysis of rotations,
in space and in space-time [95, 96].   Consider two successive rotations of a

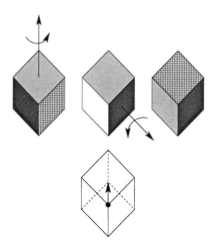

Figure 4.4: *Rotations of a coloured cube.*

cube with coloured faces, through $\pi/2$ rad, about two of the four-fold axes
of the solid body. The operation that achieves the same transformation in
a single step is obviously a rotation of $2\pi/3$ rad about the three-fold axis
along the body diagonal through the marked corner in figure 4. What is
remarkable is that this simple and obvious algorithm cannot be formulated
in terms of the law of vector combinations. It is fair to ask, if not vectors,
then what? The solution to this puzzle is purported to have struck Hamilton
with the same force as Archimedes' eureka experience. He found the formula

$$\mathbf{i}^2 = \mathbf{j}^2 = \mathbf{k}^2 = \mathbf{ijk} = -1$$

that describes the effect of a spatial rotation on a vector in terms of a hyper-complex number of unit norm, and carved the result on a bridge in Dublin. A hypercomplex number is of the form

$$a_o + a_i e_i$$

where a_o and the a_i are real numbers, whereas e_i are generalizations of $\sqrt{-1}$. If there are three such e_i, so that $i = 1, 2, 3$, the hypercomplex number is called a *quaternion*, and e_i obey rules of multiplication:

$$e_i^2 = -1 \quad , \quad i = 1, 2, 3$$

$$e_i e_j = \epsilon_{ijk} e_k \quad (i \neq j)$$

in which

$$\epsilon = \left\{ \begin{array}{c} 1 \\ -1 \\ 0 \end{array} \right\} \text{ when } i, j, k \text{ is an } \left\{ \begin{array}{c} \text{even} \\ \text{odd} \\ \text{no} \end{array} \right\}$$

permutation of $1, 2, 3$. These numbers do not obey all laws of algebra of complex numbers. They add like complex numbers, but their multiplication is not commutative. In the algebra K^4 (or H) over this field K, using Hamilton's notation, the basis elements are 1, **i**, **j**, **k** such that $\mathbf{i}^2 = \mathbf{j}^2 = \mathbf{k}^2 = -1$, $\mathbf{ij} = \mathbf{k}$, $\mathbf{ji} = -\mathbf{k}$, $\mathbf{jk} = \mathbf{i}$, $\mathbf{kj} = -\mathbf{i}$, $\mathbf{ki} = \mathbf{j}$, $\mathbf{ik} = -\mathbf{j}$, and the product of the quaternions

$$(\alpha + \beta \mathbf{i} + \gamma \mathbf{j} + \delta \mathbf{k})(\alpha - \beta \mathbf{i} - \gamma \mathbf{j} - \delta \mathbf{k}) = \alpha^2 + \beta^2 + \gamma^2 + \delta^2$$

The power of quaternions lies in the way that they describe rotations (T 1.1.9). The rotation about an axis through angle θ is represented by a quaternion

$$q = \cos \frac{\theta}{2} + \sin \frac{\theta}{2}$$

Hence $qq^* = 1$, and q is a quaternion of unit norm. If vector **r** is taken as a quaternion r with zero scalar, one finds

$$
\begin{aligned}
qrq^* &= q\{(\mathbf{r} \cdot \mathbf{n})\mathbf{n} + (\mathbf{n} \times \mathbf{r}) \times \mathbf{n}\}q^* \\
&= (\cos \frac{\theta}{2} + \sin \frac{\theta}{2}\mathbf{n})\{(\mathbf{r} \cdot \mathbf{n})\mathbf{n}\}(\cos \frac{\theta}{2} - \sin \frac{\theta}{2}) \\
&= (\mathbf{r} \cdot \mathbf{n})\mathbf{n}(1 - \cos \theta) + \mathbf{r} \cos \theta + (\mathbf{n} \times \mathbf{r}) \sin \theta \\
&= r'
\end{aligned}
$$

The effect of rotation is thus completely specified with quaternion q. For some purposes it is convenient to introduce quantities σ_j in place of e_j, by equating $\sigma_i = ie_j$, with $i = \sqrt{-1}$. With a matrix representation of complex numbers[7] and Hamilton's quaternion notation

$$\sigma_x = \begin{bmatrix} 0 & 1 \\ 1 & 0 \end{bmatrix} = i\mathbf{i}, \qquad \sigma_y = \begin{bmatrix} 0 & -i \\ i & 0 \end{bmatrix} = i\mathbf{j}, \qquad \sigma_z = \begin{bmatrix} 1 & 0 \\ 0 & -1 \end{bmatrix} = i\mathbf{k}$$

$$\sigma_x^2 = \sigma_y^2 = \sigma_z^2 = 1$$

$$\sigma_x\sigma_y = -\sigma_y\sigma_x = i\sigma_z \text{ (and cyclic permutations)}$$

According to this matrix formulation the quaternion, also known as a rotation operator or spinor transformation, becomes

$$R = \cos\left(\frac{\theta}{2}\right)\mathbf{I} - i\sin\left(\frac{\theta}{2}\right)(\sigma_x\cos\alpha + \sigma_y\cos\beta + \sigma_z\cos\gamma) \qquad (4.7)$$

with α, β and γ angles between rotation and Cartesian axes.

Returning to the solution of a coloured cube the single rotation R_3, equivalent to a product of two rotations by $\pi/2$ rad about the $z(R_1)$ and $x(R_2)$ axes, follows directly as

$$
\begin{aligned}
R_3 &= R_2 R_1 \\
&= \frac{1}{\sqrt{2}}(1 - i\sigma_z) \cdot \frac{1}{\sqrt{2}}(1 - i\sigma_x) \\
&= \frac{1}{2}(1 - i\sigma_x + i\sigma_y - i\sigma_z) \\
&= \cos 60° - i\sin 60°(\sigma_x - \sigma_y + \sigma_z)\frac{1}{\sqrt{3}}
\end{aligned}
$$

According to Hamilton's rule this result implies a net rotation through $2\pi/3$ rad about a line that makes equal angles with the three axes, as already found by inspection.

All common rotations, with many engineering applications, have the same type, being rotations about an axis and known therefore as cylindrical rotation. Rotation of another type, less well known than cylindrical rotation, but more important as the mode that allows isolated regions of space to rotate

---

[7]All matrices of a special form

$$\begin{bmatrix} \alpha & \beta \\ -\beta & \alpha \end{bmatrix}$$

combined by matrix addition and matrix multiplication are isomorphic to the field of complex numbers.

freely in the surrounding medium, without causing entanglement, is spherical rotation. This mode defines rotation around a point. A mechanical model to demonstrate spherical rotation can be constructed by suspending a practice golf ball from flexible wires fixed onto a wooden frame, as shown in figure 5.

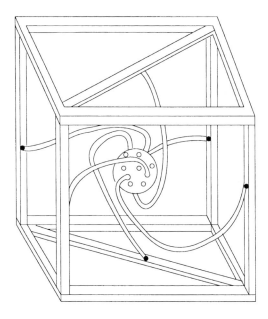

Figure 4.5: *Mechanical model [94] constructed with a practice golf ball suspended by flexible strings in a wooden framework.*

The best way to demonstrate the motion was found [94] to be starting with a rotation π rad about a horizontal axis to produce a configuration shown in figure 5. The ball can be rotated indefinitely about its vertical axis without the wires becoming permanently entangled. The initial arrangement is restored after each rotation of 4π, *i.e.* two complete revolutions. The total motion differs from normal rotation about an axis; the difference arises with the initial half twist about a horizontal axis. As the ball is then rotated about the vertical axis, the axis of the initial half turn also rotates. The composite motion is more like a continuous wobble than a rotation and the three dimensions of space therefore participate more symmetrically in the motion.

As an alternative model, imagine a cannister that is filled completely with a gelatinous medium. Imagine also that there is a small magnetized ball suspended in the jelly with adhesion to a reasonable degree at all contact surfaces. By means of an external magnetic field the ball can be inverted by

the spin axis the spinning core must turn about an axis perpendicular to the spin axis; the pertinent operator is $\begin{bmatrix} 0 & -1 \\ 1 & 0 \end{bmatrix}$. The inverted spin state is given by $\begin{bmatrix} 0 & -1 \\ 1 & 0 \end{bmatrix} \begin{bmatrix} e^{i\omega t} \\ 0 \end{bmatrix} = \begin{bmatrix} 0 \\ e^{i\omega t} \end{bmatrix}$. The inverted antispin state is $\begin{bmatrix} 0 \\ e^{-i\omega t} \end{bmatrix}$. The four spin states identified above correspond to possible states conventionally associated with an electron and its anti particle, or positron. However, uncountable intermediate spin states are allowed with this model. For instance, starting from an initial state $\begin{bmatrix} \cos \chi \\ \sin \chi \end{bmatrix}$ states between normal spin and antispin in a continuum are generated according to

$$\begin{bmatrix} e^{i\omega t} & 0 \\ 0 & e^{-i\omega t} \end{bmatrix} \begin{bmatrix} \cos \chi \\ \sin \chi \end{bmatrix} = \begin{bmatrix} e^{i\omega t} \cos \chi \\ e^{-i\omega t} \sin \chi \end{bmatrix} = \begin{bmatrix} 0 \\ e^{-i\omega t} \end{bmatrix}_{\chi=\pi/2} = \begin{bmatrix} e^{-i\omega t} \\ 0 \end{bmatrix}_{\chi=0}$$

in which 2χ is the inclination of the axis with respect to spin-up.

A spinor system such as

$$\begin{bmatrix} e^{i\omega t} & 0 \\ 0 & e^{-i\omega t} \end{bmatrix} \begin{bmatrix} \phi_1 \\ \phi_2 \end{bmatrix}$$

transformed in a Lorentzian frame has space and time derivatives [94] that, combined with the operators of $SU(2)$, generate a differential equation equivalent to Dirac's equation. With essentially the same method it will be shown here that with the Galilean invariant form it produces Schrödinger's equation, also in its linear form that more clearly defines electron spin. Not only is the Galilean transformation much simpler, but it also serves to emphasize the point that electron spin is not a relativistic effect. One starts with the spinor system as it appears to an observer moving along the x-coordinate, i.e.

$$\begin{bmatrix} \phi_1 e^{i(\omega t - kx)} & 0 \\ 0 & \phi_2 e^{-i(\omega t - kx)} \end{bmatrix} = \begin{bmatrix} \phi_1 e^+ & 0 \\ 0 & \phi_2 e^- \end{bmatrix},$$

in shorthand notation.

Forming the derivatives of the spinor

$$\frac{\partial \phi}{\partial t} = \frac{\partial}{\partial t} \begin{bmatrix} \phi_1 e^+ \\ \phi_2 e^- \end{bmatrix} = \begin{bmatrix} i\omega \phi_1 e^+ \\ -i\omega \phi_2 e^- \end{bmatrix} = i\omega \begin{bmatrix} \phi_1 e^+ \\ -\phi_2 e^- \end{bmatrix}$$

$$\frac{\partial \phi}{\partial x} = i \begin{bmatrix} -k\phi_1 e^+ \\ k\phi_2 e^- \end{bmatrix} = -ik \begin{bmatrix} \phi_1 e^+ \\ -\phi_2 e^- \end{bmatrix}$$

$$\frac{\partial^2 \phi}{\partial x^2} = k^2 \begin{bmatrix} \phi_1 e^+ \\ -\phi_2 e^- \end{bmatrix}$$

It follows that

$$\frac{1}{i\omega}\frac{\partial\phi}{\partial t} = \frac{1}{k^2}\frac{\partial^2\phi}{\partial x^2}$$

with similar results for y and z. In three dimensions

$$-i\frac{\partial\phi}{\partial t} = \frac{\omega}{k^2}\nabla^2\phi$$

This expression resembles Schrödinger's equation

$$-i\hbar\frac{\partial\Psi}{\partial t} = \frac{\hbar^2}{2m}\nabla^2\Psi$$

The two equations are identical for $k = 2\pi/\lambda$, $m = (\hbar k)^2/2\omega$, $k^2/\omega = 2m/\hbar^2$, i.e. the De Broglie condition, $p = h/\lambda$.

To turn the squared Schrödinger operator

$$S^2 = -i\hbar\frac{\partial}{\partial t} - \frac{\hbar^2}{2m}\nabla^2 \quad , \quad S^2\Psi = 0$$

into a spinor operator one writes it in a linear form

$$S = -i\hbar A\frac{\partial}{\partial t} + \frac{i\hbar\nabla}{2m} + C \quad , \quad S\Phi(spinor) = 0,$$

using matrices that form a basis for the algebra of $SU(2)$. In this way

$$S = A\begin{bmatrix} 1 & 0 \\ 0 & 1 \end{bmatrix}\frac{\partial}{i\partial t} - \begin{bmatrix} 0 & i \\ i & 0 \end{bmatrix}\frac{\partial}{\partial x} - \begin{bmatrix} 0 & -1 \\ 1 & 0 \end{bmatrix}\frac{\partial}{\partial y} - \begin{bmatrix} i & 0 \\ 0 & -i \end{bmatrix}\frac{\partial}{\partial z} + C\begin{bmatrix} 1 & 0 \\ 0 & 1 \end{bmatrix}$$

in which the four matrices represent the identity operation and rotations by π rad about the x, y and z axes respectively. Substituting the derivatives obtained before gives

$$S\Phi = \left\{\hbar\omega A\begin{bmatrix} 1 & 0 \\ 0 & 1 \end{bmatrix} + \hbar k_x\begin{bmatrix} 0 & -1 \\ -1 & 0 \end{bmatrix} + \hbar k_y\begin{bmatrix} 0 & -i \\ i & 0 \end{bmatrix} + \hbar k_z\begin{bmatrix} -1 & 0 \\ 0 & 1 \end{bmatrix}\right.$$
$$+ \left. C\begin{bmatrix} 1 & 0 \\ 0 & 1 \end{bmatrix}\right\} \times \begin{bmatrix} \phi_1 e^+ \\ -\phi_2 e^- \end{bmatrix} \tag{4.8}$$

The three spatially linked matrices are identified as the Pauli matrices σ_i. Choosing the matrices

$$A = \begin{bmatrix} 0 & 0 \\ 1 & 0 \end{bmatrix} \quad \text{and} \quad C = \begin{bmatrix} 0 & 2m \\ 0 & 0 \end{bmatrix}$$

and expanding to 4×4 matrices, equation (10) becomes

$$S \begin{bmatrix} \psi \\ \chi \end{bmatrix} = \left\{ \begin{bmatrix} \mathbf{O} & 2m\mathbf{I} \\ \hbar\omega\mathbf{I} & \mathbf{O} \end{bmatrix} + \hbar \begin{bmatrix} \mathbf{k}\cdot\sigma_i & \mathbf{O} \\ \mathbf{O} & \mathbf{k}\cdot\sigma_i \end{bmatrix} \right\} \begin{bmatrix} \psi \\ \chi \end{bmatrix} = 0$$

$$\implies \begin{bmatrix} \hbar\mathbf{k}\cdot\sigma_i & 2m\mathbf{I} \\ \hbar\omega\mathbf{I} & \hbar\mathbf{k}\cdot\sigma_i \end{bmatrix} = 0$$

where $\mathbf{I}$ and $\mathbf{O}$ are unit and null 2×2 matrices respectively, ψ and χ are two-component objects, *i.e.* $\begin{bmatrix} \phi_1 \\ \phi_2 \end{bmatrix}$ and $\begin{bmatrix} \phi_3 \\ \phi_4 \end{bmatrix}$. Written in more familiar form, one has

$$(\sigma\cdot\mathbf{p})\psi - 2m\chi = 0$$

$$E\psi - (\sigma\cdot\mathbf{p} = 0$$

Eliminating χ leads to

$$E\psi = \frac{(\sigma\cdot\mathbf{p})^2}{2m}\psi \tag{4.9}$$

According to general properties of Pauli matrices $(\sigma\cdot\mathbf{p})^2 = p^2$; hence (9) is recognized as Schrödinger's equation, with E and p in operator form. On defining the electronic wave functions as spinors both Dirac's and Schrödinger's equations are therefore obtained as the differential equation describing respectively non-relativistic and relativistic motion of an electron with spin, which appears naturally.

The nature of spin as revealed with a model of spherical rotation is fully consistent with all known attributes of electron spin [97]. It represents an intrinsic magnetic moment of one bohr magneton[8]

$$\mu_B = \frac{e\hbar}{2mc} \, ,$$

although the electron has a spin of only $\frac{1}{2}\hbar$. The postulate of electron spin was made to account for the remarkable fact that quantum states of zero angular momentum (*e.g.* the state 1^2S of H) splits two ways in a magnetic field. Whereas classical angular momentum is a conserved quantity, the quantum analogue ($\hat{L}$) is not. The quantum-mechanically conserved quantity is

---

[8]This assumption was made by Uhlenbeck and Goudsmit, the discoverers of electron spin, and later shown [98] to be the correct value for an electron viewed as a rapidly rotating body, not anywhere exceeding c in tangential velocity.

$\hat{J} = \hat{L} + \hat{S}$, which includes the spin. This condition is reminiscent of the way in which instantaneous magnetic interaction between two like charges approaching at right angles, appears to violate Newton's third law. According to the right-hand rule[9] forces between the charges are, as shown in figure 6, equal, but not opposite.

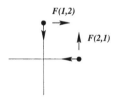

Figure 4.6: *Equal, but not opposite, magnetic forces of interaction between two identical charges approaching at right angles.*

To explain this anomaly [99] the missing momentum is deemed to be transferred to the field, and only when field momentum is added to the mechanical momentum of the charges is the total momentum conserved. Similar considerations apply to angular momentum [100]. In this instance conservation requires that the field acquire angular momentum, in addition to mechanical angular momentum that a particle might have. In the case of a quantum-mechanical particle spin must be regarded as a circulating flow of momentum density in the electron field. In this sense spin is associated not with the internal structure of an electron but rather with the structure of its wave field. The connection with the undulating wavelike region that surrounds a spherically rotating electron at half frequency is immediately obvious.

The conservation of angular momentum is a consequence of isotropy or spherical rotational symmetry of space (1.3.1). An alternative statement of a conservation law is in terms of a nonobservable, which in this case is an absolute direction in space. Whenever an absolute direction is observed, conservation no longer holds, and *vice versa*. The alignment of spin, that allows of no intermediate orientations, defines such a direction with respect to conservation of angular momentum. One infers that space is not rotationally symmetrical at the quantum level.

---

[9]This innocent rule manifests the chirality of space-time, one of its most fundamental features which is consistently ignored; it is the key to the mysterious quantized orientation of spin.

4.8   The Real Electron

Three fundamental aspects of the electron have been considered from three points of view. To rationalize the mass the electron was considered to be a standing wave trapped in a phase-locked cavity. To prevent the wave from dispersing it was defined to be a wave packet (soliton) of two components with phase velocities respectively less than and exceeding the speed of light. In order to understand the charge of the electron it was necessary to examine the tachyonic wave component more closely. Its superluminal speed was found to be due to motion in a fifth dimension along a geodesic of the compacted klein coordinate. The trajectory of a moving charged particle describes a spiral. The inertial motion occurs along the geodesic of non-Euclidean four-dimensional space-time, accompanied by, or locked to, an orbiting component with a sense of rotation that determines the sign of the charge. The soliton itself freely rotates in spherical mode, causing some disturbance of the surrounding medium. This 'propwash' is observed as the spin of the electron.

It is desirable to link all essential features of the electron to a single unifying concept at a deeper level of understanding. Many models have been proposed, and a common theme is that space-time or the aether, viewed as a continuous featureless fluid, might, when distorted become the source of all observable matter and energy. Such a fluid is expected in principle to behave according to the equations of hydrodynamics for steady flow, or their quantum-mechanical analoque, which was shown by Madelung [31] to be just the Schrödinger equation. By substituting a wave function of the form $\psi = \alpha e^{i\beta}$ into

$$\frac{hi}{2\pi}\frac{\partial\psi}{\partial t} = \left(\frac{h^2}{8\pi^2 m}\nabla^2 - V\right)\psi$$

the equation separates into two parts:

$$\frac{h}{2\pi}\frac{\partial\beta}{\partial t} = \frac{h^2}{8\pi^2 m}(\nabla\beta)^2 + V - \frac{h^2\nabla^2\alpha}{8\pi^2 m\alpha} \tag{4.10}$$

$$\frac{\partial\alpha}{\partial t} = \frac{h}{4\pi m}\left[2\nabla\alpha\nabla\beta + \alpha\nabla^2\beta\right] \tag{4.11}$$

By substituting $\beta = -(2\pi m\phi)/h$, these equations reduce to

$$\frac{\partial\alpha^2}{\partial t} + \nabla\cdot\left(\alpha^2\nabla\phi\right) = 0 \tag{4.12}$$

$$\frac{\partial\phi}{\partial t} + \frac{1}{2}(\nabla\phi)^2 + \frac{V}{m} - \frac{h^2\nabla^2\alpha}{8\pi^2 m^2\alpha} = 0 \tag{4.13}$$

In this form there is a close correspondence with the equations of hydro-dynamics, or vortex-free flow of a fluid under the influence of conservative forces. Equation (12) resembles a hydrodynamic continuity equation if α^2 is considered to be a density and if the stream velocity $v = \nabla\phi$. As $\nabla \times v = 0$,

$$\frac{dv}{dt} = -\frac{\nabla V}{m} + \nabla \left(\frac{h^2 \nabla \alpha}{8\pi^2 m^2 \alpha} \right)$$

The first term on the right is interpreted as a force density and the second corresponds to a quantity like $-\int \frac{dp}{\rho}$, with pressure p and mass density ρ, or a function of an internal stress gradient rather than actual stress, as in hydrodynamics.

In a refined form of the theory [37] the same quantity features as the well known quantum potential. The notion of a particle emerges in this theory in the form of a highly localized inhomogeneity that moves with the local fluid velocity, $\mathbf{v}(\mathbf{x}, t)$, thus as a stable dynamic structure that exists in the fluid, for example, as a small stable vortex or a pulse-like distortion. To explain why the causal theory needs probability densities it is argued [37] that the Madelung fluid must experience more or less random fluctuations in its motion to account for irregular turbulence. The turbulence necessitates a wave theory to describe the motion of vortices embedded in the fluid. The particle velocity is therefore not exactly $\nabla S/m$, nor is the density exactly $|\Psi|^2$.

Having established the equivalence of a hydrodynamic model with the Bohm interpretation, the entire electronic theory outlined above becomes re-duced to a common basis on demonstrating unitary correspondence between the de Broglie-Bohm theory and the dual-wave formulation of Corben and others [86], based on Corben's stipulation that two components of a dual wave are space-like and time-like respectively; *i.e.* a tachyon and a bradyon. Following de Broglie this condition is achieved by substituting the factorized form of the wave function

$$\Psi(x^\mu) = R(x^\mu) \exp\left[iS(x^\mu)/\hbar\right] \tag{4.14}$$

into the time-like Klein-Gordon equation for a particle of rest mass M_o,

$$\left[\Box + (M_o c/\hbar)^2\right] \Psi = 0 \tag{4.15}$$

After separating real and imaginary parts in the usual way, two equations are obtained:

$$\left[\partial_o S(x^\mu)\right]^2 - \left[\nabla S(x^\mu)\right]^2 = M_o^2 c^2 + \frac{\hbar^2 \Box R(x^\mu)}{R(x^\mu)}$$

$$\partial_o R(x^\mu)\partial_o S(x^\mu) - \nabla R(x^\mu)\nabla S(x^\mu) = -\frac{1}{2}R(x^\mu)\Box S(x^\mu)$$

The first is a relativistic Hamilton-Jacobi (HJ) equation for a particle with variable rest mass [101]

$$m_o = \sqrt{M_o^2 + \frac{\hbar^2}{c^2}\frac{\Box R(x^\mu)}{r(x^\mu)}} \tag{4.16}$$

In the expression for energy

$$E = \frac{M_o c^2}{\sqrt{1-\beta^2}}$$

this means that if M_o becomes imaginary, the particle is allowed to move faster than light with energy becoming neither infinite nor imaginary. In non-relativistic approximation, the quantum potential

$$V_q(x^\mu) = \frac{\hbar^2}{2M_o}\frac{\Box R(x^\mu)}{r(x^\mu)}$$

reduces to the form derived from the Schrödinger equation.

To demonstrate the equivalence with Corben's construct the factorized function (14) is assumed to be a superposition of amplitudal and exponential waves - one space-like and the other time-like, $i.e.$

$$\left[\Box + (m_o c/\hbar^2)\right]\exp\left[(iS(x^\mu)/\hbar\right] = 0$$

$$\left[\Box - (m_o' c/\hbar)^2\right]R(x^\mu) = 0 \tag{4.17}$$

and their superposition to be a solution of the Klein-Gordon equation (15). Hence from (16)

$$M_o = \sqrt{m_o^2 - \frac{\hbar^2}{c^2}\frac{\Box R(x^\mu)}{R(x^\mu)}} = \sqrt{m_o^2 - (m_o')^2}$$

in agreement with Corben's theory. In the case of a time-independent space-like field $R(x^\alpha)$, the wave equation (17) reduces to

$$\left[\nabla^2 + (m_o' c/\hbar)^2\right]R(x^\alpha) = 0$$

and the quantum potential takes a form

$$V_q(x^\alpha) = -\frac{\hbar^2\nabla^2 R(x^\alpha)}{2M_o R(x^\alpha)} = \frac{(m_o' c)^2}{2M_o}$$

Thus carriers of the field $R(x^\mu)$ and the source of quantum potential can be space-like particles.

The previous conclusion immediately clarifies the mystery of non-local interaction through the space-like nature of the quantum potential field. All theories actually agree that superluminal motion occurs in the interior of the electron as first discovered by Dirac, but a non-local connection is not restricted to the interior of an electron: it can occur in any region of high quantum potential, for instance in the interior of an atom or a small molecule. As the quantum potential is inversely proportional to mass, non-local interaction within more complex and more massive bodies becomes less significant. External classical potentials also have a disruptive influence on non-local interaction; claims that such connections exist over galactic distances might be inflated, but within the domain of chemical reactions they must be of decisive importance.

Attempts to formulate a causal description of electron spin have not been completely successful. Two approaches were to model the motion on either a rigid sphere with the Pauli equation [102] as basis, or a point particle using Dirac's equation, which is pursued here no further. The methodology is nevertheless of interest and consistent with the spherical rotation model. The basic problem is to formulate a wave function in polar form $\Psi = Re^{iS/\hbar}$ as a spinor, by expressing each complex component in spinor form

$$\Psi^a = R^a e^{iS^a/\hbar}, \qquad a = 1, 2$$

in which R^a and S^a are real functions. Alternatively

$$\Psi^a = R^a e^{iS/\hbar} \varphi^a$$

in which R and S are real amplitude and phase functions and

$$\varphi^a = r^a e^{is^a/\hbar}$$

is a two-component object satisfying conditions $\varphi^\star \varphi = r^\star r = 1$, $\frac{1}{2}(s^1 + s^2) = 0$, so that it has two real degrees of freedom.

Rotation about a point is described in terms of Euler angles. Rotation of the Cartesian axes (XYZ) into $(X'Y'Z')$ is accomplished in terms of three clockwise rotations: (i) through an angle ϕ about Z into (X_1Y_1Z); (ii) by an angle θ about X_1 into (X_1Y_2Z') ; (iii) by χ about Z' into $(X'Y'Z')$. These

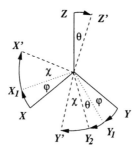

Figure 4.7: *Diagram to define Euler angles.*

three rotations, in terms of eq. (9) are

$$(i) \qquad \cos\left(\frac{\phi}{2}\right) + i\sin\left(\frac{\phi}{2}\right)\sigma_3 = \left(\begin{array}{cc} e^{i\phi/2} & 0 \\ 0 & e^{-i\phi/2} \end{array}\right)$$

$$(ii) \qquad \cos\left(\frac{\theta}{2}\right) + i\sin\left(\frac{\theta}{2}\right)\sigma_1 = \left(\begin{array}{cc} \cos\left(\frac{\theta}{2}\right) & i\sin\left(\frac{\theta}{2}\right) \\ i\sin\left(\frac{\theta}{2}\right) & \cos\left(\frac{\theta}{2}\right) \end{array}\right)$$

$$(iii) \qquad \cos\left(\frac{\chi}{2}\right) + i\sin\left(\frac{\chi}{2}\right)\sigma_3 = \left(\begin{array}{cc} e^{i\chi/2} & 0 \\ 0 & e^{-i\chi/2} \end{array}\right)$$

and the resultant transformation

$$(i)(ii)(iii) = \left[\begin{array}{cc} ie^{i(\chi+\phi)/2}\cos\left(\frac{\theta}{2}\right) & ie^{i(\phi-\chi)/2}\sin\left(\frac{\theta}{2}\right) \\ ie^{i(\chi-\phi)/2}\sin\left(\frac{\theta}{2}\right) & ie^{-i(\phi+\chi)/2}\cos\left(\frac{\theta}{2}\right) \end{array}\right] \qquad (4.18)$$

The matrix of equation (18) is of a special type $\left[\begin{array}{cc} \alpha+i\delta & -\gamma+i\beta \\ \gamma+i\beta & \alpha-i\delta \end{array}\right]$ that represents a quaternion, and its determinant $\alpha^2 + \beta^2 + \gamma^2 + \delta^2 = 1$. Its rows and columns are interpreted as components of unit spinors. A spinor of magnitude $(\Psi^\star\Psi)^{\frac{1}{2}} = \rho^{\frac{1}{2}} = R$ may therefore be written as

$$\Psi = Re^{i\chi/2}\left[\begin{array}{c} \cos\left(\frac{\theta}{2}\right)e^{i\phi/2} \\ i\sin\left(\frac{\theta}{2}\right)e^{-i\phi/2} \end{array}\right] = Re^{iS/\hbar}\varphi$$

The Hamilton-Jacobi phase is thereby identified as the angle of rotation of a rigid body about an axis in the direction of the spin.

When this form is substituted into the Pauli equation it separates (like Schrödinger's equation) into a continuity equation and a HJ equation

$$\frac{\hbar}{2}\left[\frac{\partial\chi}{\partial t} + \cos\theta\frac{\partial\phi}{\partial t} + \frac{1}{2}mv^2 + V_q + V_s + \frac{2\mu}{\hbar}\mathbf{B}\cdot\mathbf{S} + e\mathbf{A}_o + V\right] = 0$$

for spin vector $\mathbf{S} = \frac{\hbar}{2}(\sin\theta\sin\phi, \sin\theta\cos\phi, \cos\theta)$. The HJ equation differs from the classical equation for a spinning particle in having two extra terms, a quantum potential V_q and a spin-dependent quantum potential $V_s = -\frac{\hbar^2}{8m}\left[(\nabla\theta)^2 + \sin^2\theta(\nabla\phi)^2\right]$. The precession of the spin vector is described by

$$\frac{d\mathbf{s}}{dt} = \mathbf{T} + \frac{2\mu}{\hbar}\mathbf{B}\times\mathbf{S}$$

in which $\mathbf{T}$ is an additional quantum torque that produces an extra non-classical rotation of a spin that precesses in a classical field, and eventually it aligns the spin at $\theta = 0$ or π rad with respect to the field direction. In this sense spin is a manifestation of the interaction between an electron and the vacuum, which here represents the medium that carries the quantum potential and its associated field. In the absence of an external field ($\mathbf{B} = 0$) the spin vector still precesses under the influence of quantum torque that seeks a special direction in space-time. The most important unanswered question about electron spin relates to the meaning of this special direction. Only under the influence of an applied field is this spin orientation projected into three-dimensional space.

4.9 Electronic Interactions

The accepted mode of interaction between a pair of electrons involves exchange of photons. Until this exchange has been logically formulated, no model of an electron can be considered adequate. As in the case of an electron there is a conflict between wave and particle models, and as before, it may be necessary to reject both points of view as too simplistic and to seek an alternative model of the photon that reflects all known properties, including wave- and particle-like behaviour. The key to the problem lies in the nature of interaction as an exchange, which implies equal participation of the emitter and absorber. Useful ideas in this direction have been formulated by several authors.

When calculating the direct interaction of an electron with the total environment, without using a field concept, Tetrode [103] found that all interactions along the entire worldline (past and future) must be taken into account. This observation led to the important conclusion that any interaction, over whatever distance, must involve both partners in a symmetric exchange. Emission therefore never happens unless an absorber is lined up and ready to interchange energy with the emitter. The process is highlighted with the example of a star, at a distance of 100 light years, emitting light to reach an observer a hundred years later, taking off before the birth of the latter. Although the observer becomes physically involved only a hundred years after emission, he still contributed, by some response, to initiation of the exchange one hundred years previously. There is no field involved and the radiation energy appears to be lost for the duration only to reappear subsequently at completion of the interaction. The rate of emission should not be deemed to be determined by the amount of matter in the universe as two competing receptors might well interfere with each other.

A somewhat different perspective was offered by Lewis [104] on the premise "that an atom never emits light except to another atom, and that it is absurd to think of light emitted by one atom regardless of the existence of a receiving atom as it would be to think of an atom absorbing light without the existence of light to be absorbed". This proposal replaces the idea of mere emission by the idea of transmission, or a process of exchange of energy between two definite atoms or molecules. The process is postulated to be perfectly symmetric so that one atom is no longer the active agent and the other an accidental passive recipient, but both atoms play symmetric parts in the process of exchange. Prior communication that is essential to establish the course of exchange is noted by Lewis to be "repugnant to all our notions of causality and temporal sequence". An answer was found in the geometry of space-time and special relativity. A two-dimensional diagram is used to distinguish between space-like (OX) and time-like (OT) events, along the singular lines (OL) and (OL'), that belong to neither class, as shown in figure 8. The locus of a material particle occurs invariably along a time-like

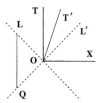

Figure 4.8: *Two dimensional diagram of four-dimensional space-time to distinguish between time-like and space-like events.*

line (like OT') and the slope of this line with respect to OT represents the velocity of the particle. The limiting line OL' represents the velocity of light; it belongs to neither sector and the radiative process represents no motion at all.

If two lines OT and QL represent the space-time loci of two material particles, the intercepts OL and OQ of singular lines between these loci have mathematically and physically zero length. Two atoms with such loci are therefore in virtual contact at any two points such as O and L or O and Q, which are connected by singular lines. Two atoms in *virtual contact*, like two atoms in actual contact, hence have no problem to establish whether their relationship allows the transfer of energy, even though the time of emission and time of absorption may be separated by thousands of years. This statement depends on an arbitrary choice of time axis. If OT moves closer to OL not only the time elapsing, but also the spatial distance between O and L approach zero. This thinking is consistent with the notion of non-local interaction that occurs in the theory of Bohm. Communication of this type does not proceed along the normal time-like channels with a maximum speed of transmission c.

The space-time symmetry underlying the Lewis model requires further analysis. It has often been speculated that the known universe is one of a pair of symmetry-related worlds. Naan argued forcefully [105] that an element of PCT (Parity-Charge conjugation-Time inversion) symmetry within the universal structure is indispensible to ensure existence. The implication is co-existence of material and anti-material worlds in an unspecified symmetric arrangement. Hence any interaction in the material world must be mirrored in the anti-world and it will be shown that this accords with the suggested mechanism of interaction.

Consider an elementary source such as a vibrating electron producing electromagnetic radiation that propagates through space and time. Its PCT mirror image is an absorber transmitting what is perceived to be negative energy in a negative time direction in the anti-world. The total interaction between emitter and absorber, and its mirror event in anti-space, are shown in figure 9. An interaction defined in this way corresponds exactly to the absorber theory of Wheeler and Feynman [106]. According to this theory an accelerated charge in otherwise charge-free space does not radiate electromagnetic energy unless acted on by a field arising from other particles. These fields are represented by one-half the retarded plus one-half the advanced solutions of Maxwell's wave equation. The law of force is symmetric with respect to past and future. It is assumed that there are sufficiently many particles to absorb completely the radiation emitted from the source. The consequences of this model have been considered in detail by Cramer [78, 107] and the formalism was shown to be, not only equivalent to standard quantum field theory but also able to account for non-local interactions of the Einstein-Podolsky-Rosen (EPR) [3] type, commonly described in terms of Bell's [40] inequality. In simple words the interaction amounts to a handshake between emitter and absorber by means of a pair of retarded and advanced waves. These are the $+t$ and $-t$ solutions of equation (2). The advanced wave that retrogresses in time is emitted by the absorber when the retarded wave from

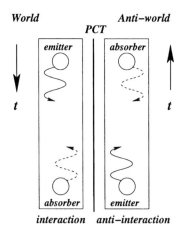

Figure 4.9: *Interactions in the world and anti-world.*

the emitter arrives, and it reaches the emitter at the exact moment when the retarded wave is first emitted. The net effect is the transfer of one quantum of energy from emitter to absorber. The interaction is confined to the period between times of emission of the retarded (t_1) and advanced (t_2) waves and to the space between emitter and absorber. It can be viewed as a standing wave that exists for the period $t_2 - t_1$ between emitter and absorber, and is also known as a photon. Cramer distinguishes between emitters of two types, compared graphically in the Minkowski diagrams, figure 10.   Each

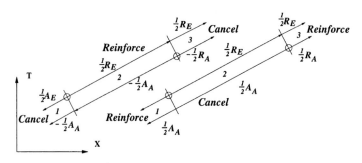

Figure 4.10: *Two modes of interaction between emitter and absorber.*

emitter produces half-amplitude retarded and advanced waves $\frac{1}{2}R_e$, $\frac{1}{2}A_e$; the absorber produces $\frac{1}{2}R_a$, $\frac{1}{2}A_a$. In interaction of type I the responding wave from the absorber is exactly π rad out of phase with the incident radiation and these waves reinforce in region (2) between emitter and absorber, and cancel elsewhere. An observer viewing this process perceives no advanced radiation but describes the event as emission of a full-amplitude retarded wave by the emitter, with appropriate energy loss and recoil, followed by absorption of this wave by an absorber at some later time, with energy gain and recoil. In interaction of type II the signs of waves produced by the absorber are reversed, with the result that the waves cancel in region (2) and reinforce elsewhere. Whereas interaction of type I describes the transmission of a photon there is no absorption in the event of type II. The emitted radiation in this case can be argued to provide a model for the neutrino. Cramer models non-local interaction at a distance as the vector sum of advanced and retarded waves involving two absorbers and a single emitter resulting in a space-like connection between the absorbers as shown in figure 11. An interaction of type I is possible only if the emitter and absorber have disparate energies. Most potential emitters in the universe are at the same vacuum

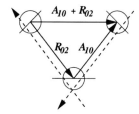

Figure 4.11: *Minkowski diagram to explain non-local space-like communication between two absorbers.*

energy level and they find no absorbers with which to establish an interaction. Their emitted waves simply interfere amongst themselves to establish what is commonly known as the radiation field, which is in universal equilibrium. The wave emitted by an electron is therefore balanced by the residual radiation from all other sources in the universe which constitute a resultant wave that converges on the electron. The interference between this pair of waves was shown by Wolff [84] to have a local maximum in the immediate vicinity of the electron and to decay like an inverse radial (Coulomb) function through space, as in equation (3). The resulting standing wave in the radiation field is not equivalent to the photon identified before and corresponds to the virtual photons of field theory.

The electron and its charge can hence be identified with a local maximum in the radiation field, embedded in a sea of virtual photons. The amplitude Φ corresponds to the electric potential of the electron and significantly does not become infinite as $r \to 0$, but approaches Φ_o. The interference between divergent and convergent waves therefore achieves the same, and more, as renormalization in field theory.

4.10   Chemical Aspects

Beyond a general recognition of their importance there is no consistent theory to model the participation of electrons in chemical processes. The well-known empirical procedures to describe charge transfer within and between molecules are commonly introduced with the disclaimer that a full quantum-mechanical treatment would be required to describe the mechanisms more rigorously. However, attempts to formulate such a treatment are rare.

The empirical models are of two kinds. The course of organic reaction mechanisms is mapped out by curved arrows that represent the transfer of electron pairs. Electrochemical processes, on the other hand are always analyzed in terms of single electron transfers. There is a non-trivial difference involving electron spin, between the two models. An electron pair has no spin and behaves like a boson, for instance in the theory of superconductivity. An electron is a fermion. The theoretical mobilities of bosons and fermions are fundamentally different and so is their distribution in quantized potential fields.

In reaction theory electrons are often considered to be delocalized while electrochemically they are either treated as point particles, or as waves, when invoking tunnel effects. Without agreement on the size and shape of an electron there hence seems to be unspoken acceptance that both of these factors are variable. This conclusion may not be too far off the mark, but still there must be some principle that regulates the degree of flexibility. Logically it should depend on the immediate environment and not on the need to support a favourite point of view. Sufficient guidelines have emerged in this review to formulate a working model that describes the size and shape of an electron in different environments.

The size-and-shape factor is of special importance with respect to electron confinement in atoms, molecules, crystals and interfaces. This confinement, empirically characterised by parameters such as electronegativity is the decisive fundamental factor that decides chemical reactivity. The demonstration that atomic electronegativity is equivalent to the chemical quantum potential of the valence state [108] holds the key to molecule formation by electron pairing and space-like delocalization. It opens a new angle on the nature of chemical binding, molecular structure, chemical equilibrium and surfaces.

Chapter 5

Chemical Concepts

5.1 Introduction

Bohmian theory, also known as the causal interpretation of quantum theory, assumes that classical and non-classical concepts merge in the quantum limit, defined as the point where the quantum potential becomes insignificant. In this formalism particle properties are associated with the wavefronts of wave mechanics, and concepts such as trajectory, space coordinate, velocity and angular momentum acquire concrete meaning also within quantum theory. The difference between classical and quantum systems relates to the emergence of the concepts quantum potential and quantum torque that enrich the interpretation by supplementing the classical ideas of kinetic energy and angular momentum. In the science of chemistry there are many ideas such as electronegativity, valence state, chemical potential and molecular structure that appear to have both classical and quantum attributes. It is a logical next step to look for a reinterpretation of chemistry in terms of Bohmian theory. An immediate result is the plausible account of the stability of matter in terms of stationary states of zero kinetic energy. This result explains, in turn, the nature of the valence state defined as the state of atomic activation due to environmental factors, at which an electron decouples from the atomic core. The decoupled electron only has quantum potential energy which allows non-local interaction with an activated neighbourhood. The formation of molecules results from the disruption of the non-local quantum potential field. Compared with the interaction within the molecule, interaction with distant objects becomes rather feeble as the total holistic wave function of the activated medium partitions into a product state representing distinct molecules. The quantum potential energy supersedes the classical concepts of electronegativity and chemical potential, whereby chemical reactivity and equilibrium conditions can in principle be quantified.

Molecular stationary states, like the atomic, are states of close-to-zero kinetic energy and angular momentum of motion. The guiding principle that promotes chemical binding is the quenching of angular momentum. As a vector quantity orbital angular momentum of motion responds to its three-dimensional environment and dictates specific conformation of molecules in terms of nuclear positions. Optical activity is the result of residual angular momentum of motion that cannot quench in chiral molecules. The resultant orbital magnetic moment interacts with the polarized photon field. In some cases the quenching of angular momentum depends critically on the anti parallel alignment of planes of circulating charge and for this reason the resultant structure resists twisting out of this plane to constitute a barrier to rotation. Many traditional concepts of theoretical chemistry are incompatible with the causal model. This includes the concept of π-bonding that often requires the promotion of an s-electron into a p-state for which it lacks the necessary angular momentum. Some predicted structures resemble their classical analogues, but many others have unexpected non-classical bonding patterns.

5.2   The Chemical Problem

Chemical reactions are initiated with activation of mixtures of stable reactants by any of several mechanisms including thermal, photochemical, compressive, electrochemical and catalytic activation. It is generally agreed that activation, by whichever means, consists of preparing the system in its valence state, also known as the promotion state, although there is no consensus on the definition of this valence state.

One way to identify a valence state could begin with a simple prospective reactant, such as a neutral atom, and to study its behaviour during a controlled process of gradual energizing with an activation process. If this process is simulated with a computer, compression would be the simplest mechanism to handle. It is achieved by confinement of an atom in an impenetrable sphere of adjustable radius and calculating the electronic energy as a function of compression radius.

5.3   Compressed Atoms

The quantum-mechanical description of the hydrogen atom is of central importance as the only example from which to derive many generalizations pertaining to chemical systems. It is nevertheless important to note that it represents real chemical systems rather poorly. Its most glaring defect is perhaps complete neglect of environment as a factor that shapes chemical behaviour. The way in which the electronic wave functions of a free atom extend to infinity is clearly at variance with chemical reality, especially in those cases in which atoms find themselves in a crowded environment at the onset of a chemical reaction. In such a situation interatomic contacts consist largely of interacting negative charge clouds. An environment of an atom is therefore approximated well by a uniform electrostatic field that prevents electronic charge density from extending indefinitely. Hence the boundary condition $\psi \to 0$ as $r \to \infty$ might be replaced by $\lim_{r \to r_o} \psi = 0$, with $r_o \ll \infty$, to obtain a more realistic solution to atomic wave equations for chemically interacting systems.

5.3.1   Particle on a Sphere

The problem of a confined electron in the valence state is identical to that of a particle confined to a line segment, which is controversial because it does not predict the classical situation ($p = \pm \hbar k$) in a limit of large quantum number. For a barrier that is high but finite, the electron begins to move like a free particle before it reaches the classical limit and then shows the correct behaviour.

   Apart from this detail it follows that a spherically confined particle has eigenfunctions of the form

$$Y_l^{m_l}(\theta, \phi) \jmath_\ell(kr)$$

and energy

$$E = \frac{\lambda_\ell^2 h^2}{8m\alpha}$$

λ_ℓ defines the first zero at radius $a = \lambda_\ell \pi$ for spherical Bessel functions, $\jmath_\ell(x)$, $x = kr$:

$$\jmath_o(x) = \frac{\sin x}{x}$$

$$\jmath_1(x) = \frac{\sin x}{x^2} - \frac{\cos x}{x}$$

$$\jmath_2(x) = \left(\frac{3}{x^3} - \frac{1}{x}\right)\sin x - \frac{3}{x^2}\cos x$$

etc.

From this, $\lambda_o = 1$, $\lambda_1 = 1.4303$, $\lambda_2 = 1.8335$ etc.

In terms of the causal model the kinetic energy in every stationary state with $m_\ell = 0$ is zero; hence the total calculated energy is pure quantum potential energy. To confirm this, recall that V_q must be constant for the confined particle, $i.e.$

$$-\frac{h^2 \nabla^2 R}{8\pi^2 mR} = K ,$$

or

$$\nabla^2 R + \frac{8\pi^2 mKR}{h^2} = 0 \qquad (5.1)$$

Equation 1 is simply Schrödinger's amplitude equation for the confined particle, provided that

$$K = V_q = E_g = h^2/(8m\alpha^2),$$

and R is the spherical Bessel function of order zero that terminates at $\alpha = \pi$. The quantity E_g is remarkably periodic when compared for all atoms and is found to follow the same trend as an empirically derived quantity known as "electronegativity", more correctly identified as the Fermi energy of an atom in its valence state [10], $viz.$ the atomic chemical potential. This result elevates the old concept to the most fundamental property of an atom, namely its quantum potential energy.

5.3.2   Ionization Radii

Wave equations for non-hydrogenic atoms can be solved numerically with a Hartree-Fock Self-Consistent-Field (HF-SCF) method (T7.3.1) under the boundary condition, that forces the wave function to vanish at a finite distance r_o. Hydrogen, as a special case, has been examined many times [109, 76, 110, 111, 112, 113, 25], using various numerical techniques for integration of the wave equation under compression. The single most important finding was an increase of electronic energy, until the ionization limit is reached on compression to a critical radius. In the HF treatment of non-hydrogenic atoms the boundary condition is introduced on multiplying all one-electron wave functions by the step function

$$S = \exp\left[-\left(\frac{r}{r_o}\right)^p\right], \text{ with } p \gg 1$$

as part of an iterative procedure [24]. The general response is the same as for hydrogen, except that all ground-state levels are affected at the same time. The highest levels are the most sensitive, but even the deepest core levels show an increase. As for hydrogen, the valence level eventually reaches the ionization limit, but this condition cannot be interpreted directly as an ionization event, as was done for hydrogen.

The quantum numbers n and l, which describe the energy levels of particular electrons, have a precise meaning only at the beginning of the iterative procedure, when they refer to hydrogen-like one-electron wave functions. As the potential field varies during the various stages of iteration they lose this meaning and no longer represent sensible quantum numbers. They are simply retained as a book-keeping device, not to indicate that the independent-electron shell structure used as a starting configuration is preserved in the final self-consistent ground state of the atom. This condition is clearly reflected in the observation that the ionization energy of a free atom resembles only superficially the HF orbital energy of the most energetic valence electron.

Another complication is that in a standard HF calculation all electrons with the same labels n, l are assigned similar energies at the average of the multiplet for that level. One cannot reasonably expect all electrons at the valence level to reach the ionization level simultaneously on excitation of the multiplet. The energy in excess of that of the ground state must clearly be redistributed so as to promote a single electron towards the ionization limit. Some excess energy assigned to lower levels must likewise be transferred to electrons at the highest level and contribute to their excitation. This problem is reminiscent of the question of spectral transitions involving groups of states with energies within broad intervals. The resolution invokes Fermi's Golden Rule (T7.2.1) that relates transition probabilities to the densities of excited levels. Inverting the argument, but not the logic, for the excess energy in compressed atoms, one arrives at probabilities of energy transmission proportional to the intervals between any calculated HF level and the valence level. This algorithm was built into the HF-Slater software to analyze atomic compressions across the periodic table [24]. The mechanism allows transfer of energy from deeper levels to a valence electron that reaches the ionization limit at a radius characteristic of each atom.

The ionization radii calculated for all atoms with this procedure show remarkable periodicity that mirrors many trends observed or inferred empirically for atomic properties such as electronegativity, covalent radii *etc.* Also indicated is a simple explanation for promotion of atoms into their valence state, before a chemical reaction commences [114]. This generally accepted mechanism, never satisfactorily explained before, can be accounted for simply in terms of environmental pressure. Whenever an atom is crowded because of high pressure or temperature or even concentration on a catalytic surface, the valence electron becomes promoted towards its ionization limit. In this limit the atom enters the valence state as an electron becomes decoupled from the core. When this electron encounters another, similarly promoted, pairing occurs and electronic energy is decreased on the formation of a chemical bond.

This description of an electron in the valence state relates to an endless debate among theorists who interest themselves in the physics of chemical bonding. The moot point is whether bond formation is due to decrease of potential energy or of kinetic energy. Arguments on either side are equally convincing and equally inconclusive. It is not uncommon to find the same theorist on occasion defending both points of view. The most likely conclusion is that both arguments are fallacious.

It was argued that an electron reaches the valence state by decoupling from the core into a state of zero potential energy. As the total energy is also zero, so is the kinetic energy. At the same time this electron is best described as a free particle in a spherical enclosure, of radius α, and therefore with ground-state energy

$$E_g = \frac{h^2}{8m\alpha^2} \neq 0$$

Small wonder that Sommerfeld and Welker [76], when they first encountered this state for hydrogen, identified it as a situation in need of further investigation. The fact of the matter is that no logical description of this electronic state exists within conventional quantum theory. It makes sense only in the framework of the Bohm interpretation.

5.4   Molecular Cohesion

Sufficient background is here established to present a consistent picture of molecular cohesion in terms of atomic compression, interpreted according to the quantum potential. It is argued that binding can occur only if one reactant in a mixture becomes promoted to the valence state, which is characterized primarily by the quantum-potential energy of the valence electron. A complication to be addressed is the fact that various species (atoms) reach the valence state under varied conditions. The nature of a chemical interaction that occurs after promotion is shown to depend on the relative states of available reactants. Development of these ideas into procedures that predict quantitative aspects of molecular cohesion must take into account the non-local nature of quantum interactions and the relatively minor contribution of kinetic energy. If, in practice, kinetic energy is ignored altogether, a calculation of bond properties reduces to a simple electrostatic calculation.

This argument should not be interpreted to mean that the formation of a chemical bond is a classical process. It means rather that the quantum-mechanical process, as it unfolds in a chemically crowded environment, appears no more complicated than a simple electrostatic interaction. Only one valence electron per atom is assumed to participate in bond formation; this excitation is assumed to leave the monopositive core unaffected. The standard Heitler-London procedure, well known as the first successful quantum-mechanical treatment of bonding in the molecule H_2 , could therefore be used to describe the single bond between any pair or couple of atoms in terms of valence-state functions.

Any empirical observations about the course of chemical interactions and the dependence of chemical reactivity and equilibrium on factors such as electronegativity, can now be rationalized directly in terms of the more fundamental quantum-potential concept. The most important aspect of this argument is that quantum potential energy of any atom (and in principle of any molecule) can be calculated from first principles. Without any further assumptions this information can be used to predict that the qualitative course of a chemical reaction amounts to overcoming a potential barrier of known size. This procedure is discussed in the next sections with indications of how it could eventually be used to provide a quantitative prediction of reaction kinetics and mechanism. Pursuing the argument into qualitative analyses of energy factors associated with bond formation, a simple relationship between bond energy and bond length is predicted, with the important conclusion that the bond-order concept is an artefact and that all covalent bonds are of the same type.

Several approximations that allow simple estimates of bond parameters are presented as a demonstration that predictions based on quantum potentials are of correct order, and not as an alternative to well-established methods of quantum chemistry. In the same spirit it is demonstrated that the fundamental thermodynamic definition of chemical equilibrium can be derived directly from known quantum potentials. The main advantage of the quantum potential route is that it offers a logical scheme in terms of which to understand the physics of chemical binding. It is only with respect to electron-density distributions in bonds that its predictions deviate from conventional interpretations in a way that can be tested experimentally.

5.5   The Valence State

The key to molecular cohesion and chemical binding is the valence state - a truly non-classical state of matter. It arises from excitation of a valence electron to the point at which it decouples from the parent atom but remains associated with it because of environmental confinement. It can be likened to a microparticle on a line, constantly reflected off the walls and eventually resembling an isotropic distribution of waves the velocity of which fluctuates about a zero mean. At this point the particle has little kinetic energy, and most energy is in the quantum potential.

The physical state of the valence electron is not obvious. In a formal sense an amount of negative charge equivalent to one electron is delocalized through the volume of a hollow sphere, although it could be argued that the core region should be inaccessible. The latter condition would envisage a valence electron to be spread across a spherical shell between the core and an enclosing surface. In fact there is no need to decide whether the valence electron is smeared over the surface of the limiting sphere or across the entire volume: the external electrostatic effects of the two charge distributions would in fact be identical .

Like its energy, the angular momentum of the valence electron also becomes spherically averaged. It rotates in spherical mode and its total angular momentum appears as quantum torque. Like a spinning electron this orbital quantum torque sets up a half frequency wave field that resonates non-locally with the environment. The most general solution to the wave equation of a spherically confined particle therefore is the Fourier transform of the spherical Bessel function

$$ J_o(kr) = \frac{\sin(kr)}{kr} $$

which has an infinite number of roots. The required transform (T 3.9) is the box function

$$ b(r) = \begin{cases} \sqrt{2\pi}/2\alpha & , \quad |r| < \alpha \\ 0 & , \quad |r| > \alpha \end{cases} $$

As a working model the electron can be characterized by a real wave function that terminates at the ionization radius and has a uniform amplitude throughout the sphere, *i.e.* according to the step function [115]

$$ \psi(r) = \left(\frac{3c}{4\pi n} \right)^{\frac{1}{2}} \left(\frac{1}{\alpha} \right) \exp\left[-\left(\frac{r}{\alpha} \right)^p \right] \ , \ p \gg 1 \qquad (5.2) $$

In this expression c is a proportionality factor to exclude the core and n is the principal quantum number that corresponds to the highest occupied energy level of the free atom. Any simulation of interactions in the valence state might begin with this formulation.

Next, imagine that the promoted atom is one of many, all similarly activated by a static field of applied pressure. All atoms are in the same valence state and interact non-locally through quantum torque and the quantum-potential field, which becomes a function of all particle coordinates. This condition implies a holistic interaction including all atoms and corresponding to the formation of a metal. The cohesion that derives from quantum interaction of all delocalized valence electrons suffices to keep nuclei in localized positions generated by the applied field, even after the field is removed.

In terms of this analysis the nuclear arrangement is established through the motion of atoms under influence of an applied field. To discover how nuclear connectivity relates to quantum effects one would have to solve Schrödinger's equation for the total ensemble, featuring all electrons and nuclei in the Hamiltonian. Only then would the role of nuclear quantum potentials, guiding the atoms into an optimal distribution, become evident. Once electronic and nuclear motions are separated[1] however, this information is destroyed.

5.6 Electronegativity

Of all the many concepts developed in the past to provide a theoretical foundation for Chemistry, electronegativity is one of the more enduring. This age-old concept was developed and refined over time as a measure of chemical affinity to differentiate between chemically antagonistic qualities: earth and air, base and acid, metal and non-metal, electropositive and electronegative, but always in a qualitative sense. It still features in many theories, albeit like the concepts valence, bond and structure [2], without "first-principle underpinning". This means that they are not represented by quantum-mechanical observables, have served a noble purpose in the past and are now obsolete. Primas [2] cautions against this, stating the task of theoretical chemistry to sharpen and explain chemical concepts and not to reject a whole area of inquiry. It is in this spirit that electronegativity will be re-examined here and shown to be an intrinsic property of an atom, promoted into its valence state by environmental pressure.

The first quantitative definition, due to Pauling [117], described electronegativity as:

The power of an atom in a molecule to attract electrons to itself.

On this basis a numerical scale, based on thermochemical data and designed to account for the increased strength of covalent bonds between unlike atoms,

---

[1]In terms of the Born-Oppenheimer approximation [116].

5.6.1   Ontological Interpretation

The intuitive concept of electronegativity can be defined more clearly in terms
of the Bohmian valence state.

Chemical reaction occurs between reactants in their valence state, which is
different from the ground state. It requires excitation by the environment, to
the point where a valence electron is decoupled from the atomic or molecular
core and set free to establish new liaisons, particularly with other itinerant
electrons, likewise decoupled from their cores [114]. The energy required
to promote atoms into their valence state has been studied before [24] in
terms of the simplest conceivable model of environmental pressure, namely
uniform isotropic compression. This was simulated by an atomic Hartree-
Fock procedure, subject to the boundary condition that confines all electron
density to within an impenetrable sphere of adjustable finite radius.

The immediate effect of atomic compression is a uniform elevation of
electronic energy levels, leading to a mixing of states and resonance transfer of
excess energy to higher levels. The valence state is reached at a characteristic
compression radius (α) when an electron has been promoted to the level of
zero energy. This is the ionization limit at which the electron is decoupled
from the core and finds itself in a hollow sphere. The energy of this valence
electron is that of a free particle so confined and depends only on the critical
compression radius. Its ground-state energy is

$$E_g = \frac{h^2}{8m\alpha^2}$$

and since α is characteristic of each atom, characteristic energies are predicted for atomic valence-state electrons. It is the atomic equivalent of the
Fermi energy of an electron at the surface of the Fermi sea in condensed
phases, and in that sense represents the chemical potential of the valence
electron for each atom. Electronegativity is defined within density functional
theory in almost identical terms [124].

Relating E_g to electronegativity provides a theoretical basis of this concept. Here is a definition without assumptions and derived directly from first
principles. It is a function of the electronic configuration of atoms only and
emerges naturally as the response of an atom to its environment. It is indeed
the tendency of an atom to interact with electrons [117] and the fundamental
parameter that quantifies chemical affinity and bond polarity.

5.6.2   Absolute Electronegativities

Calculation of electronegativities [108] formed part of an investigation to
obtain multiple ionization radii of atoms for comparison with a range of

empirical quantities like ionic radii, compressibilities, polarizabilities, met-
allization parameters and others, with and without relativistic corrections.
This necessitated streamlining of the modified Herman and Skillman FOR-
TRAN code [125] used before [24]. The heart of the program remains the
Hartree-Fock-Slater algorithm with the step-function modification [24] that
allows adjustment of the boundary condition, but with all code converted
into C. A root finder based on the shape of the one-particle valence-electron
energy curves [24] was added. This improved the accuracy and speed of finding ionization radii at selected values of the p-parameter in the step function
[24],

$$S = \exp\left[-\left(\frac{r}{r_o}\right)^p\right]$$

The major difference between the new and previously reported values [24] is
the use of $p = 100$ rather than 20, and relativistic corrections in all cases.

Ground-state confinement energies based on first ionization radii are listed
in the Appendix to chapter 5. The listed values are seen to mirror the known
trends of various electronegativity scales. The scale for comparison represents Pauling electronegativities, according to Nagle [126]. To match the
Pauling scale it is necessary to define electronegativity in terms of confinement energies as $\chi = \sqrt{E_g}$. These values are also shown in the appendix.
The discrepancies between the calculated and Pauling electronegativities on
the same scale are recorded graphically in figure 1. The observed scatter

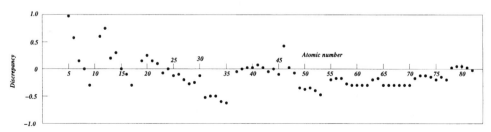

Figure 5.1: *Discrepancies between calculated and Pauling electronegativities,*
$\chi_{cal} - \chi_P$

is mainly due to variations between different periodic families. This variation could easily be eliminated by the introduction of some group-related
parameter, but without gain of new conceptual insight.

Comparison of ionization radii with a variety of empirical radii has shown
[127] that the HFS compression which cannot handle hydrogen at all and
gives inconclusive results for helium, also does not model lithium and beryl-
lium too well. Apart from these four elements, the closed-sphere free-electron

energies calculated by this procedure reflect the expected periodic trends in atomic valence states, fermi energies, and electronegativities well.

Confinement energies are related to the Configurational Energies of Allen, featured as a third dimension of the Periodic Table [128]. Configurational energies are obtained from multiplet averaged high-resolution spectroscopic data on free atoms and regarded as the parameters required to characterize spherically symmetrical, generic one-electron potentials, from which approximate molecular and solid-state wavefunctions can be constructed [129]. It is of interest that Allen [130] highlights the correlation of this χ_{spec} with orbital energy data, which like confinement energy, are obtained by HFS calculation. In terms of Allen's prescription

$$\chi_{spec} = \frac{m\varepsilon_p + n\varepsilon_s}{m + n}$$

where m and n are the numbers of p and s electrons respectively. The one-electron energies, ε_p and ε_s are the multiplet-averaged total energy differences between a ground-state neutral atom and a singly-ionized atom. This is clearly an empirical approximation to the valence-state energies calculated here from first principles.

5.7 Chemical Equilibrium

Chemical change consists of driving an adjustment to the material composition of a system by energy changes, or conversely, it represents energy production by changing the material composition. These interdependent factors can therefore be linked by a single equation that defines their inverse relationship. Identification of atomic quantum potential with electronegativity, defined as an atomic Fermi level, or chemical potential of the valence state, opens the possibility of formulating this fundamental relationship that governs chemical interaction, from first principles.

It is assumed that the tendency of a molecular mixture to interact can be analyzed as a function of the chemical (quantum) potential energy field and some action variable that reflects mass ratios or amounts of substance. Spontaneous chemical change occurs as the chemical potential of a system decreases, i.e. while $\Delta\mu < 0$, and ceases when $\Delta\mu = 0$, at equilibrium. The quantity here denoted by $\Delta\mu$, also known as the affinity, α of the system, is the sum over all molecules, reactants and products

$$\alpha = \Delta\mu = \left(\sum_i n_i\mu_i\right)_{products} - \left(\sum_j n_j\mu_j\right)_{reactants}$$

where the $\mu_{i,j}$ are the current quantum potentials of the different kinds of molecule and $n_{i,j}$ are the numbers of each kind present at the given stage of interaction. The assumption is that an increment in action is proportional to the action itself and a linear homogeneous function of the affinity:

$$dA = \beta A d\alpha$$

$$\frac{dA}{A} = \beta d\alpha$$

$$\int_i^f d(\ln A) = \beta \int_i^f d\alpha$$

for integration between arbitrary initial and final conditions. In general

$$\ln\left(\frac{A_f}{A_i}\right) = \beta\left(\Delta\mu_f - \Delta\mu_i\right) \tag{5.3}$$

The action represents the mass ratio in terms of individual activities, *i.e.*

$$A = \prod_{i,j}\left\{\frac{(a_i)^{n_i}_{products}}{(a_j)^{nj}_{reactants}}\right\}$$

To use the relationship of equation (3) in practice a logical procedure would be to define an initial state in terms of valence-state electronegativities. However, this procedure does not provide a fixed reference point for the potential function, since the valence state for each species occurs under its own characteristic conditions.

It has been common practice to define a standard state at an action of unity, $A_i = 1$, at which point by definition $\Delta\mu_i \equiv \Delta\mu^\ominus$. Equation (3) by reference to this standard state becomes

$$\ln A_f = \beta\left(\Delta\mu_f - \Delta\mu^\ominus\right)$$

i.e.

$$\Delta\mu_f = \Delta\mu^\ominus + \frac{1}{\beta}\ln Q \tag{5.4}$$

where the reaction quotient, Q represents action relative to the standard state. For a single molecule

$$\mu = \mu^\ominus + \frac{1}{\beta}\ln a \tag{5.5}$$

Equations (4) and (5) are well-known thermodynamic expressions. A reaction attains equilibrium when the affinity becomes zero, hence

$$Q = \exp(-\beta\Delta\mu^\ominus) \equiv K$$

and K is called the equilibrium constant.

The inverse argument shows how a chemical potential field imposes a particular distribution on units of matter. Consider energy levels at μ_i and μ_j in this field and assume these to be occupied by n_i and n_j molecules respectively. These numbers can be thought of as representing the relative activities at the two levels. The balance between action and affinity is then described, as before, by the relationship

$$\ln\left(\frac{A_f}{A_i}\right) = -\beta\left(\Delta\mu_f - \Delta\mu_i\right) \tag{5.6}$$

The negative sign anticipates the fact that the number of particles at a level decreases as the potential increases.

Let a total of N molecules be spread over all available energy levels. If those molecules at the levels of interest are viewed as products produced by a reaction in progress and all others as reactants, the four related variables in equation (6) follow as

$$A_f = \frac{n_j}{\prod_{k\neq j}^{N} n_k} \quad , \quad A_i = \frac{n_i}{\prod_{l\neq i}^{N} n_l}$$

$$\Delta\mu_f = \mu_j - \sum_{k\neq j}^{N}\mu_k \quad , \quad \Delta\mu_i = \mu_i - \sum_{l\neq i}^{N}\mu_l$$

It follows that

$$\left(\frac{A_f}{A_i}\right) = \frac{n_j}{n_i}\cdot\frac{\prod_{l\neq i}^{N} n_l}{\prod_{k\neq j}^{N} n_k} = \left(\frac{n_j}{ni}\right)^2$$

$$\Delta\mu_f - \Delta\mu_i = \mu_j - \mu_i - \sum_{k\neq j}^{N}\mu_k + \sum_{l\neq i}^{N}\mu_l$$

$$= 2\left(\mu_j - \mu_i\right)$$

Hence

$$2\ln\left(\frac{n_j}{n_i}\right) = -2\beta\left(\mu_j - \mu_i\right)$$

which rearranges to

$$n_j = n_i\exp[-\beta\left(\mu_j - \mu_i\right)]$$

recognized as the Boltzmann distribution for $\beta = \frac{1}{kT}$.

The total number of molecules at all energy levels

$$N = \sum_i n_i = n_o\sum\exp[-\left(\mu_i - \mu_o\right)/kT]$$

where n_o, μ_o refer to the ground level. Hence

$$\frac{n_i}{N} = \frac{n_o \exp[-\mu_i/kT] \exp[\mu_o/kT]}{n_o \left(\sum_{i=0}^{\infty} \exp[-\mu_i/kT]\right) \exp(\mu_o/kT)}$$

$$= \frac{\exp(-\mu_i/kT)}{z}$$

where z is the partition function of statistical thermodynamics (T8.21), arrived at here without statistics or probabilities.

5.7.1 Entropy

The physics behind the previous result has some interesting consequences. When a molecule (atom) is in the valence state, as defined before in terms of critical isotropic compression, its valence electrons have quantum potential energy only and no kinetic energy. The quantum potential energy is responsible for non-local interaction within the reaction neighbourhood, but unless the valence electrons acquire kinetic energy no transformation of the activated system can usually happen. As the reaction spreads some potential energy must be converted into kinetic energy, partly to drive the dispersal of electrons and partly to work against environmental factors that inhibit the dispersal of reactants. Dispersal in the vacuum is commonly assumed to require no work, but if there is any resistance some energy will be dissipated in the form of heat and be lost to the system. The total change in energy, recognising the fact that some fraction of the heat of reaction ($\Delta\mu$), is reversibly released into the surroundings, will then be given by

$$\Delta u = \Delta\mu + q - w$$

Formal resemblance of this expression to the more familiar thermodynamic form (T8.1)

$$dU = d(\mu n) + TdS - pdV$$

is rather compelling and suggests an unexpected physical interpretation of entropy.

To better appreciate this extraordinary suggestion, consider once more a homogeneous atomic sample slowly compressed to its valence state and maintained at that pressure until it spontaneously changes into a metal while releasing an amount of chemical energy corresponding to

$$\Delta\mu = \mu_{met} - \mu_{val}$$

per atom. The lowering in potential energy arises from the redistribution of valence electrons with respect to the nuclei, *i.e.* from an atomic to a metallic stationary state. It requires no energy of dispersal and generates no entropy. For all other types of reaction electron transfer over macroscopic distances is needed. This flow of charge is different from intramolecular non-local interaction and is a source of entropy[2]. The outcome of the flow is a more disordered arrangement. In this sense the conventional interpretation seems to confuse cause and effect by stating that disorder amounts to entropy. Maybe it's the energy dissipated in the process that leads to the disorder that generates entropy. Energy dispersal may of course be mediated by other entities, apart from electrons, and also experience resistance to their motion through space. Photons are known to suffer from this as evidenced by their red shifts, or loss of energy when travelling over galactic distances. It is interesting to speculate that all quantum-mechanical particles experience friction when moving through the vacuum. It is this universal vacuum friction that renders processes irreversible and creates the arrow of time. There is no mechanism to avoid frictional dissipation of energy and reversibly return to an initial state.

The unexpected but not unreasonable conclusion is that entropy is a quantum effect, which may account for the fact that no reasonable physical interpretation of this concept has ever been advanced. If quantum and classical particles are differentiated on the basis of their mass it's not difficult to understand why only quantum particles are significantly affected in their motion through the vacuum. An important feature of this suggestion is that reversibility is here linked to a microprocess, generally assumed to be fully time-reversible and not affected by the second law of thermodynamics, largely because individual quantum processes have no mechanism. Within the ontological interpretation however, an electron that sets out from a state of high quantum potential and gathers kinetic energy as it moves through space towards some chemical encounter, is not unlike Eddington's stone[131]:

> When a stone falls it acquires kinetic energy, and the amount of the energy is just that which would be required to lift the stone back to its original height. But if the stone hits an obstacle its kinetic energy is converted into heat-energy. There is still the same quantity of energy, but even if we could scrape it together and put it through an engine we could not lift the stone back with it. Looking microscopically at

---

[2]Entropy in this sense is different from the thermodynamic state function S, which has a large reversible component, for instance as defined at phase transitions. An entropy change in the system is compensated for by an almost equal, but opposite change in the surroundings.

the falling stone we see an enormous multitude of molecules moving downwards with equal and parallel velocities. We have to notice two things, the *energy* and the *organization of the energy*. To return to its original height the stone must preserve both of them. When the stone falls on a sufficiently elastic surface the motion may be reversed without destroying the organization. Each molecule is turned backwards and the whole array retires in good order to the starting point. But what usually happens at the impact is that the molecules suffer more or less random collisions and rebound in all directions. They no longer conspire to make progress in any one direction; they have lost their organization. Afterwards they continue to collide with one another and keep changing their directions of motion, but they never again find a common purpose.

As the electron is accelerated from its stationary state, gaining kinetic energy and interacting with its environment, some energy is irradiated and lost, because of vacuum resistance. Like Eddington's stone the electron therefore irretrievably looses energy to its environment, in this case the radiation field. An inconclusive debate on this issue has been raging for many years. Since the effects are minor they are commonly ignored in the assumption that the laws of quantum mechanics are fully time-reversible. This creates the paradox of an irreversible reality based on time-reversible mechanics.

According to elementary electrodynamics an accelerated charge radiates energy, but for a free electron the mechanism of the process is problematical. The nature of the confusion and details of the debate have been well documented by Grandy[132]. It revolves around the lack of consensus about the charge, mass and structure of the electron. These quantities are all characterized differently in terms of classical, relativistic and quantum field theories. The aspects that make good physical sense in one theory cause insurmountable problems in another. In Grandy's words " an adequate theory of elementary particles remains to be found ". Although there is general agreement about the reality of *radiation damping*, also called *radiative reaction*, which deals with the resistance experienced by accelerated charges in the vacuum, there is total confusion over its origin and nature[3]. In the most advanced

---

[3] A qualitative criterion to assess the extent of radiation damping may be derived[55] from the Larmor formula that estimates the energy radiated in terms of an electronic acceleration of magnitude a, for a period of time T,

$$E_{rad} \sim \frac{2e^2 a^2 T}{3c^3}$$

treatments radiation damping is associated with nonlocal time behaviour that violates causality in the equations of motion. Another unresolved question concerns the conservation of energy and the origin of the energy which the accelerated electron radiates. A quantity of energy contained in the *Schott term*, called *acceleration energy*, which may be positive or negative, cannot be accounted for.

In terms of ontological ideas it is attractive to relate the acceleration energy as deriving from quantum potential energy in a local exchange of energy between electrons and the space medium. The time asymmetry is also less puzzling if it means dissipation of energy in a mechanism that generates entropy on a microscale[4]. This process need not be confined to accelerated charged particles but could operate whenever a (small) quantum-mechanical particle moves through the vacuum. It is important to note that the energy exchange is possible in both directions. This means that slower particles can absorb excess energy transferred to the space medium by more energetic ones. Within a closed system of many interacting particles (*e.g.* an ideal gas.) energy exchange with the vacuum introduces a random element that eventually establishes equilibrium. It also ensures that all particle trajectories are not time-reversible and so creates the arrow of time. More massive, classical bodies are not affected in the same way and the vacuum has a negligible effect on the motion. They appear to have time-reversible trajectories and only contribute entropy through friction with the atmosphere or other massive bodies. This conjecture solves the irreversibility paradox by postu-

---

If $E_{rad} \geq E_o$, the energy of the electron, the effects of radiation damping will be appreciable. In the present instance the electron starts from rest and is acted upon by an applied force for an interval T. E_o therefore is the kinetic energy after acceleration, $E_o \sim m(aT)^2$. The criterion then becomes

$$m(aT)^2 \sim \frac{2}{3}\frac{e^2 a^2 T}{c^3} \qquad \text{or} \qquad T = \frac{2}{3}\frac{e^2}{mc^3}$$

which defines a characteristic time $\tau = 2e^2/3mc^3$. The conclusion is that for acceleration times T, long compared to τ, radiation effects will appreciably modify the motion of the electron. For an electron $\tau = 6.26 \times 10^{-24}$ sec, the time taken for light to travel 10^{-13} cm. This is the sort of time involved in a radiative transition between atomic energy levels (quantum jump) for which an electron trajectory cannot even be contemplated. The timescale for electronic acceleration from the stationary valence state is more leisurely, but it clearly must generate some radiation and some perturbation of the electron's trajectory. These effects collectively contribute to the entropy and irreversibility of the reaction.

[4]No record of any discussion of radiation damping based on the ontological interpretation could be found. One major simplification would be that in terms of this approach electrons in real stationary states have no kinetic energy and will be accelerated from rest.

lating that both classical and quantum mechanics are idealized constructs that only apply in the limit of frictionless motion, that never occurs in reality. The inevitable energy loss that accompanies all motion, accumulates in the vacuum and adds a random element that renders processes irreversible and shows up as increased entropy, usually identified with disorder.

5.8 Chemical Reaction

The quantum-mechanical formulation of the progress of a reaction such as

$$A + B \rightarrow C + D$$

appears [133] to start from a stationary product state $\psi_A.\psi_B$ of mixed reactants and proceeds via the entangled valence state ψ_{ABCD} towards the final product state $\psi_C.\psi_D$. There is no obvious mechanism for such events in terms of traditional quantum theory.

In Bohmian formalism it may be argued that the reaction system, considered closed, is described at all times by an equation $H\Psi = E\Psi$ in the time-dependent wave function $\Psi(A, B, C, D)$. The product states $\psi_A.\psi_B$ and $\psi_C.\psi_D$, as well as the valence state are special solutions of this same equation under different boundary conditions. All rearrangements and transformations that determine the final outcome happen in the valence state. This valence state is not unique and is conditioned by thermodynamic factors. Under stormy conditions the reactants may be fragmented into smaller units, $A \rightarrow na_i$, $B \rightarrow mb_j$, etc. The number and nature of possible reaction products will depend critically on the degree of fragmentation. Fragmentation itself is brought about by electronic, vibrational and rotational transitions, with rates linked to the ambient conditions. The valence state may therefore be formulated in terms of variables, characteristic of either molecular fragments, atoms, nuclei, electrons and/or photons. It may be either holistic or partially holistic, with matching quantum potential. The extent of non-local interaction depends on the quantum potential and may be a factor limiting the extent of possible intramolecular rearrangement during chemical reaction. The traditional argument does not contemplate instantaneous transitions between states and intramolecular rearrangement becomes a complete mystery.

Real chemical reactions are violent affairs and quite unlikely to proceed as smoothly as the ideal metalization outlined before. Translational kinetic energy associated with reacting units is generally involved and only a fraction of activating encounters lead to binding. The final product is unlikely to incorporate all atoms promoted into the valence state; smaller fragments are

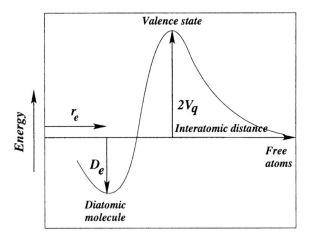

Figure 5.2: *Schematic drawing of promotion to the valence state and forma-tion of a diatomic molecule.*

The next approximation is to clamp the nuclei at classically variable coordi-nates. This approximation still allows freedom to study the electron density quantum-mechanically. However, in view of the nature of the valence state developed here there is precious little to gain by attempting all-electron cal-culations.

5.9.1   Electrostatic Model

Only two valence electrons per bond and the two monopositive cores may hence be assumed to contribute to the bonding interaction. Because the two atoms enter into reaction from stationary states and end up in a new "diatomic" stationary state, kinetic energy can be of only peripheral impor-tance, unless states of non-zero orbital angular momentum are involved. If the interaction between two uniform electronic charge clouds and two point nuclei is parametrized in terms of electrostatic point charges, the only pa-rameter seen to affect the interaction involving atoms of known ionization radii is the internuclear distance d. The point charges are defined in terms of the separate atomic volumes and the volume of overlap, V_{12} in figure 3. Any observed interatomic distance represents a balance between electrostatic attraction and the total intramolecular quantum potential. The bond energy then represents the classical electrostatic potential energy, exactly as in the case of atomic hydrogen discussed before,

$$E = K\left[\delta\epsilon\left(\frac{1}{b} + \frac{1}{d+b} - \frac{1}{d} - \frac{1}{p}\right) + X\right]$$

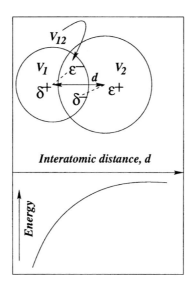

Figure 5.3: *Schematic diagram to demonstrate the electrostatic point-charge calculation of bond dissociation energy.*

K is a dimensional constant. $X = 0$, unless $\alpha_2 > \alpha_1 + d$, when $X = [(1 - \alpha_1)/(\alpha_2 - d)]^2/d$, $p = \alpha_1 + \alpha_2 - d$, $b = (\alpha_1/\alpha_2)d/2$ for $\alpha_1 < \alpha_2$.

$$\delta\epsilon = \left(\frac{1}{\alpha_2}\right)^3 \left[\frac{d^3}{16} - \frac{3d}{8}(\alpha_1^2 + \alpha_2^2) - \frac{3}{16d}(\alpha_2^2 - \alpha_1^2)^2 + \frac{1}{2}(\alpha_1^3 + \alpha_2^3)\right]^2$$

This model has been tested exhaustively for a large variety of bonds and found to provide a good description of bond strength as a function of bond length.

The energy expression for homonuclear bonds ($\alpha_1 = \alpha_2$) in atomic units, reduces to

$$E = \frac{1}{2}\left(\frac{d}{2}\right)^5 + \frac{1}{2}\left(\frac{d}{4}\right)^4 - 23\left(\frac{d}{4}\right)^3 + \frac{11}{2}\left(\frac{d}{4}\right)^2 + +\frac{25d}{16} - 5 + 3d^{-1} \qquad (5.8)$$

shown graphically in figure 3. This relationship between bond strength and bond length has been empirically inferred before[135] but never predicted by any other model of chemical bonding. In the quantum-potential model it follows from the condition that requires balance of quantum and classical potentials for stationary states. Since the classical potential is a function of the environment the bond length is allowed to vary continuously. A covalent bond is thus predicted to depend mainly on the interaction between the primary pair of valence electrons, the immediate intramolecular environment

and steric interactions. The concept of bond order is undefined and the
theory predicts that the electron density within so-called single, double, and
triple bonds between a given pair of atoms should be invariant. In the present
scheme these bonds are no more than single bonds of different strengths in
different environments.

5.9.2   The Heitler-London Method

The process of bond formation is poorly modeled with a point-charge description that contains the vital information about the relative valence states of
dissimilar atoms in terms of ionization radii only. The predicted course of
reaction is outlined in figure 4. Atomic promotion is once more modeled [136]

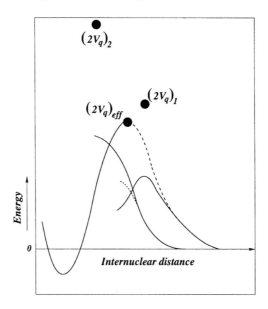

Figure 5.4: *Potential barriers and charge transfer of importance in the formation of a heteronuclear diatomic molecule.*

with isotropic compression of atoms of both types[5].

   The more electropositive atom (at the lower quantum potential) reaches
its valence state first and valence density starts to migrate from the parent
core and transfers to an atom of the second kind, still below its valence state.

---

[5]A more likely real event of promotion would be an energetic collision between the two
atoms.

The partially charged atom is more readily compressible to its promotion state, as shown by the dotted line in figure 4. When this modified atom of the second kind reaches its valence state two-way delocalization occurs and an electron-pair bond is established as before. It is notable how the effective activation barrier is lowered with respect to $2V_{qi}$, the barrier for reaction of homonuclear pair i. The effective reaction profile is the sum of the two promotion curves of atoms 1 and 2, with charge transfer.[6]

To incorporate these qualitative ideas into a quantitative account of covalent binding the Heitler-London (HL) prescription is most effective. It must be formulated in terms of two monopositive cores at coordinates $\mathbf{x}_A$ and $\mathbf{x}_B$, and two electrons of which the coordinates are represented by $\mathbf{r}_1$ and $\mathbf{r}_2$. The Coulombic interaction between these particles is expressed as

$$V = e^2 \left[\frac{1}{|\mathbf{x}_A - \mathbf{x}_B|} + \frac{1}{|\mathbf{r}_1 - \mathbf{r}_2|} - \frac{1}{|\mathbf{r}_1 - \mathbf{x}_A|} - \frac{1}{|\mathbf{r}_1 - \mathbf{x}_B|} - \frac{1}{|\mathbf{r}_2 - \mathbf{x}_A|} - \frac{1}{|\mathbf{r}_2 - \mathbf{x}_B|} \right]$$

In the HL treatment of H_2 it is assumed that the electronic wave function for the system at large internuclear separations is the product of two unmodified electronic wave functions ($1s$) of a free H-atom in its ground state, $i.e.$

$$\Phi_I = \psi(\mathbf{r}_1 - \mathbf{x}_A)\psi(\mathbf{r}_2 - \mathbf{x}_B)$$

Another possibility is

$$\Phi_{II} = \psi(\mathbf{r}_2 - \mathbf{x}_A)\psi(\mathbf{r}_1 - \mathbf{x}_B)$$

in which the two electrons are interchanged between the nuclei. As the nuclei move together the wave function is assumed to develop into a linear combination of Φ_I and Φ_{II}. The symmetry of the problem suggests equal sharing of the electrons between identical atoms according to

$$\Psi_\pm = \frac{1}{\sqrt{2}} (\Phi_I \pm \Phi_{II})$$

Recall that Φ_I and Φ_{II} are real functions because ψ represents the lowest s-states, whereby, using (3.5)

$$R^2 = \frac{1}{2} \left(\Phi_I^2 + \Phi_{II}^2 \pm 2\Phi_I\Phi_{II} \right) \tag{5.9}$$

---

[6]An essentially equivalent mechanism was proposed by Sanderson [137] in terms of electronegativity equalization.

where

$$\epsilon_{aa} = \int \psi_a(1)\psi_a(1)(1/R_{b1})d\tau$$

$$\epsilon_{bb} = \int \psi_b(1)\psi_b(1)(1/R_{a1})d\tau$$

$$S = \int \psi_a(1)\psi_b(1)d\tau$$

$$\epsilon_{ab} = \int \psi_a(1)\psi_b(1)(1/R_{a1}d\tau$$

$$\epsilon_{ba} = \int \psi_a(1)\psi_b(1)(1/R_{b1}d\tau$$

To evaluate these integrals it is necessary to transform to elliptical coordinates and substitute the atomic wave functions of eqn. (2). The three unique one-electron integrals become [115]:

$$\epsilon_{aa} = K_1 \int (\mu + \nu) \exp\big\{ - 2[R(\mu + \nu)/2\alpha_a]^p\big\}d\mu d\nu$$

$$\epsilon_{ab} = K_2 \int (\mu - \nu) \exp\big\{ - [R(\mu + \nu)/2\alpha_a]^p\big\}d\mu d\nu$$

$$S = K_3 \int (\mu^2 - \nu^2) \exp\big\{ - [R(\mu + \nu)/2\alpha_a]^p\big\}d\mu d\nu$$

where

$$K_1 = 3R^2c/8n_a\alpha_a^2$$
$$K_2 = 3R^2c/8(n_an_b)^{\frac{1}{2}}\alpha_a\alpha_b$$
$$K_3 = 3R^3c/16(n_an_b)^{\frac{1}{2}}\alpha_a\alpha_b$$

Although these integrals cannot be evaluated in closed form they are well suited for numerical evaluation. To get the two-electron integrals into an equally acceptable form it is necessary to make a few simplifying approximations. Firstly it is noted that for $p \gg 1$ the two integrals become identical. This integral is simplified by first evaluating the the spherical potential [139]

$$\int \psi_a^2(1/R_{12})d\tau_1 = (4\pi/r_2)\Big[\int_0^{r_2} \psi_a^2 r_1^2 dr_1 + \int_{r_2}^\infty \psi_a^2 r_1 r_2 dr_1\Big]$$

$$= (4\pi/r_2)\Big\{(K_a/\alpha_a^2) \int_0^{r_2} r_1^2 \exp\big[- 2(r_1/\alpha_a)^p\big]dr_1$$

$$+ (K_ar_2/\alpha_a^2) \int_{r_2}^\infty r_1 \exp\big[- 2(r_1/\alpha_a)^p\big]dr_1\Big\}$$

$$= (4\pi K_a/\alpha_a^2)(I_1/r_2 + I_2)$$

where $K_a = 3c/4\pi n_a$.

Substituting $a = 2/\alpha_a^p$ and $\nu = ar_1^p$

$$I_1 = (1/p) \int_0^{ar_2^p} a^{-3/p} \nu^{(3/p-1)} \exp(-\nu) d\nu$$

$$I_2 = (1/pa^{2/p}) \int_{ar_2^p}^{\infty} \nu^{2/p-1} \exp(-\nu) d\nu$$

These integrals can be evaluated as infinite series [140]. For $p \gg 1$ only the first term of I_1 survives, and hence

$$I_1 = (1/3)r_2^3 \exp\left(-ar_2^p\right)$$

It follows that

$$\int \psi_a^2 \psi_b^2 (1/R_{12}) d\tau_1 d\tau_2 = \left(4\pi K_a/3\alpha_a^2\right) \int r_a^2 \exp\left[-2(r_a/\alpha_a)^p\right] \left\{(K_b/\alpha_b)^p\right]\right\} d\tau$$

where $K = 3c^2 R^5/\left[64 n_a n_b (\alpha_a \alpha_b)^2\right]$. Calculated values of bond lengths and dissociation energies of homonuclear single bonds formed by the p-block elements B to Cl, using the ionization radii listed in the Appendix, compare sufficiently well with known experimental values to confirm the general applicability of the method.

For heteronuclear bonds there is an imbalance in nuclear screening that can be compensated for by the simple procedure [141], writing the effective charge product

$$Z_e^2 = \left(\frac{\varphi_1}{\varphi_2}\right)^{\frac{2}{3}} < 1 \tag{5.10}$$

for bonds between p-block atoms. For diatomic hydrides of these atoms it was established empirically that

$$Z_e^2 = (k_n \varphi_p/\varphi_H)^{\frac{2}{3}}$$

with $k_2 = 0.84$ and $k_3 = 0.70$.

These predictions are borne out quantitatively in HL calculations of electron-pair bond energies and bond lengths for heteronuclear atomic pairs [114, 142]

5.9.3 Multiple Bonding

The first convincing models of chemical bonding antedate the discovery of electron spin. That is probably why the original postulate of multiple electron pairs in a bond, often required to satisfy an octet rule, persists to this

day, despite the implied violation of the exclusion principle. Since bond order relationships however, are real, the effect of surplus electrons on bond
strength and bond length needs an alternative explanation. It is not only
electron density in excess of a bonding pair (unit bond order), but also a
charge deficiency in the valence shell that has an effect on bond parameters.
This effect is a remarkably constant function of bond order, irrespective of
the chemical identity of the bond. A universal correlation that relates bond
order to the length of bonds between oxygen and second-row atoms [143] is
shown in figure 6. The functional relationship between bond order, B, bond

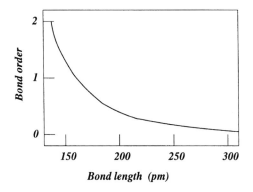

Figure 5.6: *Correlation curve between bond length and order for second row
oxygen bonds, after Brown [143]*

length, r and single-bond length, r_0 is of the type $B = (r_0/r)^n$.

The smooth relationship between bond order and bond length defines
bond order of unity between a pair of atoms with a precise ratio of 2 between
the number of available valence electrons and covalent bonds. Any deviation
from this value affects both bond order and bond strength by a fixed measure.
The primary single-bond strength is therefore assumed to be a function of
the implied regular charge distribution. Any over- or under-supply of valence
charge defines bond order different from unity and a proportional increase
or decrease in bond strength. The same argument was used by Cottrell and
Sutton [141] to model the imperfect screening of nuclei in heteronuclear diatomic molecules, by electrons not involved in the bond. They demonstrated
by model calculations the relative importance of changes in internuclear repulsion, over changes in ionic contributions, on the stabilization of molecules
with unequal centres. The functional meaning of these observations is that
changes in bond order do not modify the exchange interaction between nuclei
and bonding electrons that constitutes the single bond. The only effect of
an excess or dearth of electrons, not involved in the bonding, would then be

modification of the internuclear repulsion, by screening or descreening of the nuclei, irrespective of the supposed orbital description of the valence electrons. This assumption has been used successfully to model the strengths and bond lengths of multiple bonds up to order 4 from single-bond properties, for a variety of homonuclear and heteronuclear bonds [142, 144, 145].

Any bonding curve represents equilibrium between attractive and repulsive interactions as shown in figure 7. The best known model that describes

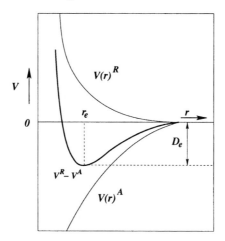

Figure 5.7: *Potential energy curve of a diatomic molecule made up of repulsive an attractive potentials.*

such an equilibrium is the Morse function [146] :

$$V(r) = D_e\{1 - \exp[-a(r - r_e)]\}^2$$

Two of its three adjustable parameters, D_e and r_e correspond to directly measurable molecular properties of bond dissociation energy and equilibrium interatomic distance respectively. The third parameter, a, is related to the force constants commonly used in spectroscopic analyses. A standard procedure [147] to obtain experimental potential energy curves is to calculate from spectroscopic data the constants k_e, g, and j that appear in the expression

$$V(r) = V_0 + \frac{1}{2}k_e(r - r_e)^2 - g(r - r_e)^3 + j(r - r_e)^4 + \dots \tag{5.11}$$

The energy reference can be chosen so that V_0 is zero and all higher terms are ignored. It is therefore possible, in principle, to obtain an exact Morse description of any single bond based on experimental data. The resulting Morse curve can then be used to generate comparative multiple-bond potential-energy curves by allowing for the effects of screening.

To establish the relationship between a and the force constants of eq. (11), the Morse function is expanded as an exponential series, using the same zero point in energy:

$$V(r) = D_e\{a^2(r - r_e)^2 - a^3(r - r_e)^3 + (7/12)a^4(r - r_e)^4 + \ldots\}$$

Near the equilibrium point $(r \approx r_e)$, all higher terms can be ignored to yield a direct relationship between a and the various force constants. In harmonic approximation

$$a_H^2 = k_e/2D_e$$

With the quadratic force constant k_e in units of mdyneÅ^{-1} and D_e in eV, the Morse constant in atomic units (a_0^{-1}) is calculated as

$$a_H = 0.9349(k_e/D_e)^{\frac{1}{2}}$$

To obtain a curve that tends to zero at infinite separation of the atoms, the Morse function is written in the form

$$V(r) = V - D_e = D_e\{\exp[-2a(r - r_e)] - 2\exp[-a(r - r_e)]\}$$

If this Morse function is used to represent any single bond, not necessarily of a diatomic molecule, the constant a calculated from the harmonic force constant may not be entirely appropriate, and especially not over the entire range of r. Before deriving multiple-bond properties from the single-bond curve it is therefore useful to optimize the Morse constant empirically to improve the match between calculated and observed single-bond values of D_e and r_e.

In practice, the single-bond curve is modified by allowing for reduced internuclear repulsion to produce potential-energy curves for bonds of higher order. The modification consists of assuming screening factors, $0 < k \leq 1$, which decrease with increasing bond order. A special feature that emerged from such modelling is the constancy of k for any bond of given order; $e.g.$ $k = 0.72$ and 0.41 for any double or triple bonds, respectively. These values suggest an approximate linear relationship between bond order b and screening factor k, of the form

$$k = -0.295b + 1.3 \tag{5.12}$$

It follows that the maximum integral bond order allowed by the scheme is 4, for which $k = 0.12$ is predicted. This predicted value was used [145] to calculate bond parameters for quadruple dimetal bonds according to the same procedure and found to reproduce known experimental values for all bond orders.

As an illustration of the procedure a set of modified Morse curves to describe C-N bonds of different order is shown in figure 8. The internuclear

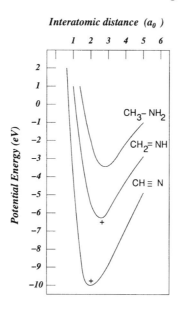

Figure 5.8: *Experimental Morse curve for* CH_3-NH_2 *and multiple-bond potential-energy curves obtained from the Morse curve by screening the internuclear repulsion [144].*

energy of repulsion over an electron-pair bond has the very simple form, $V^R = K/r$, where K is a dimensional constant. With screening, $V^R = kK/r$, k given by eqn. (12). A typical result is shown in figure 8 with experimental minima indicated by + signs. Bond properties have been calculated for C-C, N-N, O-O, S-S, C-N, C-O, N-O, C-S, Re-Re, Cr-Cr, Mo-Mo and W-W bonds. In addition the effect of reverse screening on single-bonds, adjacent to multiple bonds, were investigated in the molecules CH_3-CH=CH_2, CH_2=CH-CH=CH_2, CH_3-C≡CH, CH_2=CH-C≡CH, and CH≡C-C≡CH. Screening constants $k = 0.97$ and 0.865 were found to account for the effects of neighbouring double and triple bonds respectively.

The minima of the bonding curves of figure 8 must lie on the energy curve of figure 3, as shown in figure 9. This relationship emphasizes the fact that integral bond order is an exception rather than rule and that so-called non-bonded interaction differs from covalent bonding in degree only. On this cal universe is non-empty and therefore space is curved. The non-Euclidean nature of space-time is therefore not in question, only the extent and mode of curvature.

The original formulation of general relativity by Einstein was based on the principles of equivalence and general covariance [149]. These principles lead directly to the idea that gravitation can be explained in terms of Riemannian geometry [150]. Riemannian geometry is characterized by a symmetric tensor that general relativity theory identifies as the gravitational potential. The motion of a test particle in a gravitational field is given by a geodesic equation (T4.4).

The most general formulation of the gravitational field is in terms of gauge theory. In ordinary gauge theory a charged field is described by a complex wave function ψ, whose phase changes under gauge transformation,

$$\psi \to \psi \exp(i\alpha)$$

Since ψ depends on space-time coordinates, the relative phase factor of ψ at two different points would be completely arbitrary and accordingly, α must also be a function of space-time. To preserve invariance it is necessary to compensate the variation of the phase $\alpha(x)$ by introducing the electromagnetic potentials (T4.5). In similar vein the gravitational field appears as the compensating gauge field under Lorentz invariant local isotopic gauge transformation [150].

Chemical behaviour depends on chemical potential and electromagnetic interaction. Both of these factors depend on the local curvature of space-time, commonly identified with the vacuum. Any chemical or phase transformation is caused by an interaction that changes the symmetry of the gauge field. It is convenient to describe such events in terms of a Lagrangian density which is invariant under gauge transformation and reveals the details of the interaction as a function of the symmetry. The chemically important examples of crystal nucleation and the generation of entropy by time flow will be discussed next. The important conclusion is that in all cases, the gauge field arises from a symmetry of space-time and the nature of chemical matter and interaction reduces to a function of space-time structure.

5.10.1   Crystal Nucleation and Growth

Growing crystals is often described as art rather than science. The primary process that nucleates the growth of crystals apparently is thermodynamically forbidden and since growth after nucleation is thought to cease in the absence of dislocations, the very existence of crystals becomes something

Chapter 6

Molecular Structure

6.1 Introduction

Theoretical chemistry has two problems that remain unsolved in terms of fundamental quantum theory: the physics of chemical interaction and the theoretical basis of molecular structure. The two problems are related but commonly approached from different points of view. The molecular-structure problem has been analyzed particularly well and eloquent arguments have been advanced to show that the classical idea of molecular shape cannot be accommodated within the Hilbert-space formulation of quantum theory [161, 2, 162, 163]. As a corollary it follows that the idea of a chemical bond, with its intimate link to molecular structure, is likewise unidentified within the quantum context [164]. In essence, the problem concerns the classical features of a rigid three-dimensional arrangement of atomic nuclei in a molecule. There is no obvious way to reconcile such a classical shape with the probability densities expected to emerge from the solution of a molecular Hamiltonian problem. The complete molecular eigenstate is spherically symmetrical [165] and resists reduction to lower symmetry, even in the presence of a radiation field.

This problem is reminiscent in a sense of the difference between the traditional Copenhagen convention and the interpretation of quantum events based on the ideas of Bohm and de Broglie [166]. The Copenhagen model [167] discourages discourse in terms of classical concepts, whereas the Bohmian approach [34, 35] recognizes the merger of classical and quantum variables in the classical limit and the possibility of carrying over these concepts (*e.g.* particle, trajectory, angular momentum) from the classical to the non-classical realm. The difference between quantum and classical systems reduces to the relative importance of the quantum potential, which becomes insignificant for classical systems. It may therefore not come as a surprise if many chemical concepts that seem to border on both classical and quantum ideas find ready expression in terms of quantum potentials and the causal interpretation.

6.2   Conventional Theory

The theories and models of molecular structure and chemical bonding are closely entangled and based on a set of deceptively simple ideas first popularized by the pioneering works of Linus Pauling. The intuitive appeal of these models is such that very little effort has gone into the assessment of their scientific merit and validity although the uncritical handling of chemical concepts related to quantum theory has been clearly documented [168]. A brief critical review follows.

6.2.1   The Chemical Bond

A simple picture of chemical bonding derives from exchange interaction: Two rugby players on the run, passing the ball back and forth between them, are compelled to stay within some maximum distance from each other, as if an attractive force, mediated by the rugby ball, operates between them. The simplest possible chemical bond, in the molecule H_2^+, can be described in strict analogy, as two mobile protons that exchange a so-called valence electron between them. This interaction is called a covalent bond.

Electromagnetic interactions are thought to operate by the same mechanism, but mediated by photons. Because photons are massless they have an infinite range that defines the electromagnetic field. Unlike photons, electrons are massive particles and the covalent interaction is of shorter range, of the order 100 pm.

Covalent bonding depends on the presence of two atomic receptor sites. When the electron reaches one of these sites its behaviour, while in the vicinity of the atom, is described by an atomic wave function, such as the $\psi(1s)$, ($l = 0$), ground-state function of the H atom. Where two s-type wave functions serve to swap the valence electron the interaction is categorized as of σ type. The participating wave functions could also be of p, ($l = 1$), or d, ($l = 2$) character to form π or δ bonds respectively. The quantum number l specifies the orbital angular momentum of the valence electron. A common assumption in bonding theory is that a valence electron with zero angular momentum can be accommodated in a p or d state if a suitable s-state is not available. The reverse situation is not allowed.

To visualize the formation of chemical bonds, those parts of the atomic wave functions that extend beyond the reach of covalent interactions may be ignored as irrelevant. Depending on their mathematical structure each type of truncated atomic wave function will then be described by a geometrically characteristic boundary surface, called an orbital. Valence electrons in a bonding situation are assumed to be confined to their valence orbitals and the exchange interaction to occur only when the bonding orbitals of neighbouring atoms overlap.

Atomic wave functions with magnetic quantum number $m_l = 0$ are real functions and their corresponding orbitals can be mapped in the form of well-defined geometrical shapes. Wave functions of electrons with $m_l \neq 0$ are complex functions and do not generate orbitals in real space. But, if by some procedure, these complex functions could be transformed into real orbitals in three-dimensional space, it would in principle be possible to use these spacially directed orbitals to predict the three-dimensional shape of molecules according to the pattern of overlap. The well-known scheme of hybridization by linear combination of atomic orbitals represents such an attempt.

Starting from the degenerate p-states of equation 3.19, real wave functions, directed along three cartesian axes, are obtained first and these are used in linear combination among themselves and with an s-orbital to construct directed orbitals. As shown in figure 3.7 however, definition of real p_x and p_y orbitals by the linear combination of complex wave functions amounts to no more than a rotation of the coordinate axes. The functions p_x, p_y, p_z do hence not represent states that can co-exist on an atom and linear combinations of real functions, such as sp^2 and sp^3 have no quantum-mechanical meaning. Despite these facts, orbital hybridization remains a seductive argument because linear combinations of the real trigonometric polar functions $\cos\theta$, $\sin\theta\cos\phi$ and $\sin\theta\sin\phi$ with a spherical charge distribution, generate two-fold, trigonal and tetrahedral symmetries in one, two and three dimensions respectively. These are the classical stereochemical geometries expected for symmetrical two, three and four-coordinate molecular species.

Hybridization

In the conventional theory of chemical bonding and molecular structure the role of orbital angular momentum is ignored and the hybridization schemes based on real wave functions are claimed to predict the observed geometries of simple molecules. The method may be demonstrated by consideration of the atoms $C(s^2p^2)$, $N(s^2p^3)$ and $O(s^2p^4)$, with valence shell configurations as shown. In all cases, the linear combinations sp, sp^2 or sp^3 are assumed to lead, under suitable conditions, to linear, trigonal or tetrahedral molecular geometries, respectively. The best known example is the tetrahedral arrangement of CH_4, ascribed to four electron pairs in $4 \times (sp^3)$ hybrid orbitals, despite the fact that there are only two p-electrons in the CH_4 valence shell. It is argued that since electrons are indistinguishable any electron may occupy such a hybrid orbital, although the mathematical manipulation whereby an sp^3 hybrid is constructed cannot create the angular momentum, $L = \sqrt{l(l+1)}\hbar$ carried by a p-electron. It is necessary to assume that the density distribution of an s electron in space may be identical to that of an electron in a p state, but without the angular momentum. It follows that sp^3 hybridization defined in this way, does not imply the promotion of an electron from an s to a p-state, in a procedure such as $C(s^2p^2) \to C(s^1p_x^1p_y^1p_z^1)$, which requires a change in orbital angular momentum. Even though orbital hybridization by itself does therefore not violate the conservation of angular momentum, the electronic quantum numbers lose their meaning in the process, with the result that orbital angular momentum and electron density can no longer be specified in terms of the same variables. The assumed set of p_x, p_y, p_z orbitals does not represent the degenerate set of quantum states with $m_l = -1, 0, +1$.

Pauling's prescription [169] to specify a set of four equivalent tetrahedral bond orbitals,

$$t_{111} = \frac{1}{2}(s + p_x + p_y + p_z)$$

$$t_{1\bar{1}\bar{1}} = \frac{1}{2}(s + p_x - p_y - p_z)$$

$$t_{\bar{1}1\bar{1}} = \frac{1}{2}(s - p_x + p_y - p_z)$$

$$t_{\bar{1}\bar{1}1} = \frac{1}{2}(s - p_x - p_y + p_z)$$

in terms of the variables

$$s = 1$$
$$p_x = \sqrt{3}\sin\theta\cos\phi$$
$$p_y = \sqrt{3}\sin\theta\sin\phi$$
$$p_z = \sqrt{3}\cos\theta$$

is simply the specification of cubic body diagonals in terms of cartesian axes in spherical polar notation. This trigonometric definition of a tetrahedron has nothing to do with quantum theory. Three equivalent bond orbitals, called sp^2 hybrids, are obtained by a comparable procedure in the xy plane:

$$\frac{1}{\sqrt{3}}s + \frac{\sqrt{2}}{\sqrt{3}}p_x$$

$$\frac{1}{\sqrt{3}}s - \frac{1}{\sqrt{6}}p_x + \frac{1}{\sqrt{2}}p_y$$

$$\frac{1}{\sqrt{3}}s - \frac{1}{\sqrt{6}}p_x - \frac{1}{\sqrt{2}}p_y$$

The sp hybridization specifies positive and negative directions along a line:

$$\frac{1}{2}(s + p_x)$$

$$\frac{1}{2}(s - p_x)$$

The characterization of bonds in terms of sp, sp^2 or sp^3 orbital hybridization schemes is seen to consist of assuming either linear, trigonal or tetrahedral molecular geometry. Nothing is predicted or explained.

What hybridization schemes do achieve is to provide a mapping of the electron density in a molecule, once the three-dimensional structure in terms of nuclear positions has been specified. This is an important achievement since the chemical properties of any substance critically depend on the electron-density distribution. Since the time of Pauling's pioneering work, a lot of effort has consequently gone into the development of improved methods for calculation of the electronic structures of molecules. The many different methods, of different degrees of complexity, that have been devised, all have a number of features in common. In all cases the calculation starts by assuming a plausible geometry of the molecule. Based on the symmetry of this assumed molecular structure, a linear combination of atomic orbitals (LCAO) is used to construct a trial molecular wave function that represents the molecular electron-density function. Finally, the density is optimized mathematically as a function of the total energy as specified by the quantum-mechanical molecular Hamiltonian. Exactly as in elementary hybridization schemes, the symmetry-adapted atomic orbitals are made up of hydrogen-like radial functions, combined with real spherical harmonics[1]. The net effect is to optimize the electron density in line with the assumed molecular symmetry. When

---

[1]A minimum basis set consists of one function for each atomic orbital with distinct quantum numbers n and l.

dealing with double bonds a further decision on the σ-π classification of orbitals is required. This decision is closely related to the traditional view of multiple bonding, to be discussed next. The common claim that molecular structures can be calculated more precisely than determined experimentally will also be examined.

Multiple bonds

One of the celebrated successes of orbital hybridization is the elucidation of multiple bonding, which stems from the Lewis formulation of a chemical bond as a shared electron pair. In a compound such as ethylene the saturation of the carbon valence shells can only be achieved by the sharing of two electron pairs between the two carbon atoms.

This arrangement is forbidden by the exclusion principle. An immediate solution is by the assumption that the four bonding electrons between the carbon atoms reside in two different molecular orbitals. There are two possibilities based on sp^3 and sp^2 hybridization schemes respectively. The first possibility leads to the idea of bent bonds:

All C-H bonds are of the sp^3 type, with tetrahedral H-C-H angles. The bent bonds trace out the density along sp^3 orbitals that overlap at an angle, above and below the plane of the nuclear framework. The second possibility assumes the direct overlap of C(sp^2) hybrid orbitals. The trigonal array of three sp^2 orbitals lies in a plane. The proposed pattern of overlap between two sets of sp^2 orbitals accords well with the observed geometry of the ethylene molecule:

The same planar structure as before is obtained, now with H-C-H angles of 120°. The extra pair of valence electrons are assigned to the $2p_z$ orbitals on carbon. The sideways overlap of these orbitals is interpreted to form a π-bond.

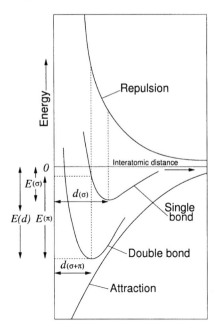

An attractive feature of the latter model is the observed values of H-C-H angles close to 120°.

The π-bonding proposal has gained general acceptance as the most convincing model to account for the planarity of ethylene and the resistance to rotation around the double bond. Any torsion of the central bond in ethylene must clearly destroy the co-alignment of the seperate p_z orbitals and thereby weaken the supposed π-overlap and hence the bond strength. An obvious anomaly is the high energy contribution associated with the indirect overlap of p-orbitals. This anomaly is put into perspective [144] in figure 1, which schematically represents the potential-energy curves of a single σ-bond and a double bond consisting of σ and π contributions. The curve for a single

Figure 6.1: *Schematic illustration of the relationship between single and double bonds*

bond represents equilibrium of repulsive and attractive interactions and has a minimum, $-E_\sigma$, at an interatomic separation d_σ. The repulsive contribution appears to increase rapidly at separations less than d_σ and at the equilibrium 'double-bond' separation the σ bond strength is rather insignificant. To account for the increased strength E_d of the double bond, it is required that

$$E_\pi >> E_d - E_\sigma$$

and as $E_d \simeq 2E_\sigma$, also that $E_\pi >> E_\sigma$. This inequality is awkward to explain in view of the generally accepted notion that σ overlap is inherently more effective than π overlap.

An alternative explanation is that the increased observed bond strength is due to an increase of σ-bond strength at the shorter internuclear distance in the double-bond[2] situation. The primary attraction is assumed to arise from the interaction between atomic cores and the σ pair in the diatomic stationary state that maximizes this attraction. Additional valence density is excluded from this state by the Pauli principle. The attraction curve of figure 1 therefore applies for any bond order. The repulsion may however, be modified by additional valence density if it screens the internuclear repulsion. Regardless of the identity of the atoms this screening must be the same for given bond order.

A strong selling point of Pauling's hybrid-orbital theory has been its rationalization of the non-steric barrier to rotation, which is characteristic of all double bonds. Since each of the sp^2 lobes in ethylene is defined to be pear-shaped there is no reason why the two CH_2 fragments should remain coplanar. However, by placing the two remaining valence electrons in the still unoccupied p_z orbitals, perpendicular to the separate CH_2 planes and assuming that these orbitals tend to stay parallel, an overall planar structure is generated. The reason for this parallel alignment of the two separate atomic p_z orbitals is obscure. The illogical common explanation in terms of sideways overlap to form a π-bond, is almost like received dogma. An amazing property of this π-overlap is that, despite the geometrical lack of common ground between the p-orbitals, it is assumed more effective than direct primary overlap.

6.2.2   The Valence Shell

All theories of molecular shape owe some debt to the idea that electronic charge accumulates in valence orbitals between atomic nuclei. The first successful method to predict the structure of molecules consisting of a central

---

[2]In this discussion bond order is interpreted strictly in terms of traditional counts of electrons per bond. There is no suggestion that bonding density changes in units of two electrons only and there is no special importance associated with integral bond orders.

atom surrounded by a number of ligand atoms, is based on this same concept. Originating with suggestions of Lewis, this valence-shell electron-pair repulsion (VSEPR) theory has been popularized largely by Gillespie [170, 171].

VSEPR

The central idea of this theory is that valence bonding pairs and lone pairs of electrons avoid each other in space because of mutual repulsion arising, either from Pauli's principle[3] or simply from electrostatic interaction. In the simplest form of the theory, valence pairs are considered as point charges on the surface of a unit sphere that surrounds the central atomic nucleus. Minimization of the energy of interaction, as a function of the electron-pair charge distribution in the spherical surface, predicts the molecular geometry by assuming bond vectors to be directed from the central atom through the point charges. The optimum geometry for 2, 3, 4, 5, 6, 7 and 8 points, calculated on this principle, is linear, trigonal planar, tetrahedral, trigonal bipyramidal, octahedral, pentagonal bipyramidal, and dodecahedral or antiprismatic, respectively.

To account for observed deviations of observed molecular geometries from the symmetrical forms, the basic theory has been refined by a number of additional assumptions, such as:

- Lone-pair densities are less concentrated than bond densities in the same valence shell and occupy more space.

- The size of a bonding pair is inversely proportional to the electronegativity of the ligand atom and directly proportional to that of the central atom.

- The extent of multiple-bond densities increases with bond order compared to a single-bond pair, although treated as a single pair.

The postulates of VSEPR theory are consistent with the partitioning of electron density according to Bader's atoms-in-molecules method [173], in which the electron pairs return as the valence shell charge concentration.

---

[3]Interpreted to mean that electrons with the same spin tend to avoid each other and therefore do not occupy the same space [172].

Bohmian Interpretation

A logical extension of the VSEPR method to yield quantitative molecular geometries requires minimization of the repulsion between properly scaled characteristic ligand point charges and lone-pair densities, at their actual radial distances from the central atom. Such modelling was performed before [174] using semi-empirical parameters to fix the relative ligand and lone-pair charges and their radial positions. In retrospect, all of these parameters can now be related to the more fundamental concepts of ionization radius and quantum potential.

The effective nuclear charge on a ligand atom was established as $Z_e = k\alpha^{\frac{1}{3}}$, where α is a characteristic radius which can now be identified with the ionization radius of the atom, and $k = 0.8$ was found empirically to produce effective nuclear charges relative to $Z(\mathrm{H}) = 1$. This formula follows from the formulation

$$\psi = \frac{1}{\sqrt{\pi}} \left(\frac{Z}{a_0} \right)^{\frac{3}{2}} \exp(-Zr/2a_0)$$

of hydrogenic radial wave functions for nuclear charge Z, i.e. $Z_e \propto \psi^{2/3}$. Using $\psi^2(r) = \phi/V = \rho_v$, the valence-electron density, it follows that $Z_e \propto (\rho_v)^{\frac{1}{3}}$. Using $\phi = c\alpha/n$, the valence density is expressed in terms of α and hence $Z_e = k\alpha^{\frac{1}{3}}$, as assumed before.

A related equation that links electronegativity to valence density

$$\chi = 0.62\rho_v^{\frac{1}{3}}$$

was obtained independently by Batsanov [175]. From the definition of electronegativity in terms of valence-state quantum potential

$$\chi = (V_q)^{\frac{1}{2}} = \frac{h}{\alpha\sqrt{8m}}$$

and the result $Z_e = (1/\alpha)\,(3c\alpha/4\pi n)^{\frac{1}{3}}$, the relationship

$$\chi = \frac{h}{\sqrt{8m}} \left(\frac{4\pi n}{3c\alpha} \right)^{\frac{1}{3}} Z_e$$

leads directly to Batsanov's equation.

Equivalent point charges to represent lone pairs were calculated as $\alpha/2$, from the ionization radius of the central atom and their effective distance from the nucleus was related to the same parameter. These values are based on the assumption that lone-pair densities should relate to valence densities and that their distance from the nucleus should depend on atomic size and oxidation state. On this basis lone pairs were identified as first-period pairs, higher-period pairs of elements in low oxidation state, like S^{-2}, and of elements in their high oxidation states. Each type assumes a characteristic value of k in $d(lp) = k\alpha$. Point charges representing ligand atoms were placed at known interatomic distances from the central atom. Equivalent point charges on ligand atoms connected by double bonds were assumed to be enhanced because of nuclear screening, as discussed in a previous section.

Minimization of electrostatic interactions as defined by the various parameters described above, was done in order to calculate bond angles in halomethanes, -silanes and -germanes, as well as one-centre compounds that contain lone pairs and double bonds. The comparison with experimental values is such as to leave no doubt about the power of the VSEPR model.

6.3 Experimental Study of Molecular Structure

Molecular structure is traditionally analyzed [176] by the methods of either diffraction, spectroscopy or *ab initio* calculation.

6.3.1 Diffraction Methods

Diffraction is almost universally accepted as the almost infallible method for molecular and/or crystal structure determination. The distinction between diffraction in the gas phase, amorphous materials, powders and single crystals is often not appreciated. The fundamental phenomenon underlying diffraction is the scatter of radiation by matter. Neutron and electron beams are considered as radiation in this context. The radiation quantum (*e.g.* X-ray) excites a unit of matter (*e.g.* electron) which on relaxation emits radiation at the same wavelength, but scattered in all directions. Scattered waves from various electrons of the same atom interfere, so that the total atomic scattering (known as the scattering factor) is a function of the radial atomic electron density distribution and the angle of scatter, with respect to the direction of incidence. The more tightly bound core electrons scatter more effectively at high angles, compared to the loosely bound valence electrons that scatter at small angles.

Scattered waves from neighbouring atoms interfere in exactly the same way and unless the atoms are ordered as in a crystal, the total diffraction pattern is a function of the radial distribution of scattering density (atoms) only. This is the mechanism whereby diffraction patterns arise during gas-phase electron diffraction, scattering by amorphous materials, and diffraction

observed data is optimized as a function of atomic positions by least squares procedures. The goodness of fit is commonly assessed in terms of the ratio,

$$R_{hkl} = \sum |F_{obs} - F_{cal}|_{hkl}/F_{obs}$$

Another factor to be taken into account is the degree of over determination, or the ratio between the number of observations and the number of variable parameters in the least-squares problem. The number of observations depends on many factors, such as the X-ray wavelength, crystal quality and size, X-ray flux, temperature and experimental details like counting time, crystal alignment and detector characteristics. The number of parameters is likewise not fixed by the size of the asymmetric unit only and can be manipulated in many ways, like adding parameters to describe complicated modes of atomic displacements from their equilibrium positions. Estimated standard deviations on derived bond parameters are obtained from the least-squares covariance matrix as a measure of internal consistency. These quantities do not relate to the absolute values of bond lengths or angles since no physical factors feature in their derivation.

The single most important factor in crystallographic analysis is crystal quality. The diffraction pattern of nearly perfect crystals, called whiskers, consists of spikes that are too sharp to measure by conventional methods. The broad peaks that are measured routinely arise from crystal imperfections, collectively called *crystal mosaicity*. The implication is that strict translational symmetry does not extend beyond small domains in a single crystal. These domains are slightly out of alignment with neighbouring domains, stacking into a three-dimensional mosaic of randomly misaligned blocks. In order to interpret the diffraction intensities it is necessary to assume that the mosaic blocks scatter independently and without interference, adding up as if all blocks had the same orientation. This theory is convenient, but not necessarily infallible. Most crystals have an arrangement somewhere between the two extremes and some measure of interference between mosaic blocks is inevitable and difficult to compensate for.

A further complication is that molecules in different blocks, or even neighbouring unit cells can have different orientations. If this happens in an orderly fashion the size of the unit cell is increased by an integral factor and a superstructure appears. Not if the alternative orientations appear at random. In that case the size of the unit cell remains the same and the different molecular images are crystallographically superimposed in the same space, with coincident atomic positions only occurring by accident.

One moiety that almost invariably appears to be disordered in crystals is the perchlorate ion. There are numerous examples in the literature, particularly because of the well-known tendency of perchlorate to promote the crystallization of complex ionic species. The perchlorate groups often occur at different sites in the same crystal with different degrees of disorder, which seem to depend on the volume of the available space between anions. In a small cavity it could be ordered, with well-defined tetrahedral geometry. The larger the void, the more severe is the disorder. It is almnost as if it has the ability to fill any free volume, within limits, to capacity, by adapting a smeared-out structure to match the shape and size of the available space. This behaviour raises the interesting possibility that a free perchlorate group is essentially structureless and assumes a tetrahedral shape only when confined into a tight spot, in line with the conjecture [178] that molecular shape is an induced property that becomes apparent only in condensed phases.

If structure and shape are not intrinsic properties of free molecules and only emerge in response to environmental pressure the interpretation of crystallographic structures becomes less obvious. The electron-density transform (1) may well be the three-dimensional projection of a four-dimensional periodic function that fluctuates with time. The possibility that crystallography looks at a time-averaged projection with the appearance of of a rigid arrangement cannot be discounted.

6.3.2 Spectroscopy

Spectroscopic measurements at virtually all accessible wavelengths have been used at some stage to obtain molecular-structure information, but many of these are now of historical interest only. The most important methods still in use for structure analysis are microwave spectroscopy, nuclear magnetic resonance spectroscopy and measurement of optical activity. All of these techniques suffer from the same disadvantage as non-single-crystal diffraction: there is no reciprocal relationship between atomic coordinates and the spectrum. Once a trial conformation has been selected however, microwave spectra translate into internal coordinates, more accurate by orders of magnitude, than crystallographic results. Another advantage of microwave spectroscopy is that it provides information on (isolated) molecules in the gas phase.

The microwave experiment studies rotational structure at a given vibrational level. The spectra are analyzed in terms of rotational models of various symmetries. The vibration of a diatomic molecule is, for instance, approximated by a Morse potential and the rotational frequencies are related to a molecular moment of inertia. For a rigid classical diatomic molecule the moment of inertia $I = \mu r^2$ and the equilibrium bond length may be calculated from the known reduced mass and the measured moment, assuming zero centrifugal distortion.

For a non-linear triatomic molecule three parameters are needed to specify (the assumed) fixed geometry, but only two moments of inertia can be measured. The situation becomes rapidly worse for larger molecules. Isotopic substitution may be used to produce more data, but the situation where sufficient data are available for a unique solution is rare. In the final analysis measured spectra may well be consistent with an assumed molecular conformation, without excluding many other classical or non-classical possibilities.

Nuclear magnetic resonance spectroscopy is the most versatile of all spectroscopic techniques and is used for the study of a wide range of materials including liquids, solids and solutions, but like other spectroscopic techniques it cannot be used independently for the direct determination of molecular structure. The technique is based on the principle that nuclei with spin have an associated magnetic moment which interacts with an applied field to produce level splittings. Radiation absorbed at the resonance frequency causes transition between the spin levels and is characterictic of the atom, its environment and the strength of the applied magnetic field. The resonance frequency is therefore sensitive to coupling with neighbouring spins through bonds, through space, and through chemical reorganization. The higher the field strength the more information the spectrum carries and the more difficult the task to disentangle the overlapping multiplets. Combined with chemical intuition the method rivals crystallography as a tool for structural analysis.

6.3.3   Theoretical Analysis

In the early days of wave mechanics there existed the reasonable expectation that a comprehensive theory of chemistry could be developed through mathematical solution of molecular wave equations. Although this expectation is no longer entertained, approximation methods based *ab initio* on quantum-mechanical concepts are widely used to optimize trial structures by energy minimization. Although such methods cannot predict molecular shape they have the important application of calculating the electron-density distribution in a molecule of known structure. Used in conjunction with X-ray crystallography it becomes feasible to address one of the basic objectives of chemistry and relate chemical effects to changes in molecular electron density.

6.3.4 Charge Density

The determination and interpretation of electronic charge distributions in molecular crystals have been reviewed by Koritsanszky and Coppens [179, 180]. Many of their formulations are used in the discussion that follows.

X-rays are scattered predominantly by electrons rather than atomic nuclei. To determine atomic coordinates, electron densities are therefore assumed to be concentrated spherically around individual nuclei. This assumption ignores all possible effects that chemical bonding may have on electronic density in molecules. Such a hypothetical array of spherical atoms located at the nuclear positions of an actual molecule in a crystal is known as a *promolecule*. Molecular structures determined by routine crystallographic methods are invariably the structures of promolecules.

To assess the effect of chemical bonding on the electron density it is assumed that the effect on core density is negligible, and that the total distortion will be due to valence-charge migration. The molecular charge density may hence be written as

$$\rho(\text{mol}) = \rho(\text{promol}) + \delta\rho(\text{val})$$

The total electron density is assumed to be a superposition of density units, each of which rigidly follows the motion of the nucleus that it is attached to. In terms of the scattering vector

$$\boldsymbol{H} = h\boldsymbol{a}_1^* + k\boldsymbol{a}_2^* + l\boldsymbol{a}_3^*$$

where $\boldsymbol{a}_i^*$ are reciprocal axes and V the volume of the unit cell, the thermally averaged *deformation density* is specified as

$$\Delta\rho(\boldsymbol{r}) = \frac{1}{V}\sum_{\boldsymbol{H}} |F_{obs}(\boldsymbol{H}) - F_{iam}(\boldsymbol{H})|\exp(-2\pi i\boldsymbol{H}\boldsymbol{r})$$

referred to the independent-atom model of the promolecule.

Deformation densities defined in this way typically show density accumulation in the bonds and lone pair regions. Exceptions were first observed [181] in standard X-ray electron-density mapping of a polycyclic molecule containing C, H, N and O atoms. A steady decrease in the order C-N > C-O > N-N > O-O, of deformation densities in bonds, was observed. The density along the O-O bond was found to be negative throughout.

In terms of hybrid-bond theory it appears reasonable that the deformation density could be negative in some bonds. When a p-block atom with n valence electrons forms a bond, the valence shell is polarized into a tetrahedral distribution with $n/4$ electrons concentrated around each potential binding site. Only one electron from each atom is required to form a bond and the predicted deformation density of $2(1 - n/4)$ electrons per bond is therefore exactly in line with the observed trend. However, polarization only happens when the bond forms and the free-atom density should be averaged, not over four potential binding sites, but over the total solid angle of 4π. It is readily demonstrated [162] that for any valence shell of less than 9 electrons the required charge concentration per bond, in each of the four tetrahedrally directed contiguous solid angles, exceeds the free-atom density. Observed negative deformation densities may therefore be symptomatic of inappropriately calculated free-atom densities, rather than deficient theory.

The most likely cause of such discrepancy is an unsuitable atomic scattering factor. That means, some factor that affects the chemical behaviour of an atom may, for instance, not be properly accounted for in the calculated electronic structure from which scattering factors are derived. The use of oriented non-spherical atomic ground-states has been proposed [182] as a possible remedy. On this basis theoretically acceptable *chemical deformation densities* have been obtained. Such usage has led to the development of aspherical-atom, or *multipole* refinement of crystallographic structures in charge-density studies.

The aspherical atomic electron density is divided into three components:

$$\rho(\boldsymbol{r}) = \rho_c + P_v\rho_v + \rho_d$$

where ρ_c and ρ_v are the spherical core and valence densities, respectivly, and the valence deformation is of the form

$$\rho_d = \sum_l R_l \sum_m P_{lm}Y_{lm}$$

in terms of a radial part and real spherical harmonics. All of these terms are calculated from Slater-type basis functions.

Deformation densities based on multipole refinements do not include the effects of thermal smearing and hence represent static densities, with a high degree of dependence on the basis sets used in generating the aspherical functions.

6.3.5   Atoms in Molecules

Crystallographic electron-density functions exhibit local maxima at the positions of atomic nuclei and, not surprisingly, several efforts to partition the density function into regions that represent individual atoms have been made.

of molecular spectroscopy [192] this is allowed for all point groups in which translational and rotational transforms are linked isomorphically. Any classical structure associated with an optically active substance must belong to the same symmetry point groups; *i.e.* they must lack any alternating axis S_n, but this restriction applies only to the chiral eigenstate and not to the total molecular Hamiltonian.

What is being argued here is that molecules have three-dimensional structures, shaped, not so much by a highly symmetrical eigenstate as by a development history that involves the fusion of quantum fragments into a partially holistic molecular framework, characterized by a factorizable eigenstate. Although Pfeifer [185] tried to resolve the issue of optical activity by coupling the molecule to the radiation field, the final objective was the same: to define the ground state as a separable product state. The model of optical activity, based on the stated assumption that chiral factors could feature in a molecular product state, may seem to simply restate the generally accepted Rosenfeld mechanism [187], but it is fundamentally different.

The conventional view is that the molecular basis of optical activity relates to an assumed non-zero scalar product of electric and magnetic transition moments that defines a rotational strength

$$R_{ab} \propto < a|\boldsymbol{\mu}|b > \cdot < b|\mathbf{m}|a >$$

in which the magnetic dipole moment operator is a complex quantity of the form

$$\mathbf{m} = \frac{eh}{4\pi imc}(\mathbf{r} \times \nabla) = -i(\mathbf{r} \times \nabla \beta_m)$$

and β_m is the Bohr magneton. Spin angular momentum is seen to make little or no contribution to chiral activity which arises mainly from orbital angular momentum. The rotational strength, as pointed out before, is subject to severe symmetry restrictions. However, while it is correct to state that a molecular eigenstate describes at the same time all possible isomers, precursors, products, conformations, oligomers and phases of a given molecule, this could apply at only one instant in the history of that molecule; when all constituent atoms are in their valence states together. At such a point the eigenstate is holistic and the pregnant system could find expression in any of the possible arrangements mentioned above. The extent of reaction that ensues, however, depends on external factors and could conceivably lead to a product with the full symmetry of the molecular eigenstate. More generally, however, intermediate fragments are produced with structures of lower symmetry and characterized by wave functions that are factors of the product eigenstate. Subsequent interactions between fragments do not proceed via the holistic valence state and local structures are established as dictated by local factors of quantum potential energy and angular momentum. The divergent properties of isomers, oligomers and the like, are due to these local factors and should not be traced back to the original holistic eigenstate. Within this scheme it becomes feasible to relate magnetic transition moments to local currents that generate electronic orbital angular momentum vectors. These vectors are not the same as the spectroscopic rotational states of a molecule.

6.4 Molecular Conformation

The concept of quantum potential has been found to elucidate the nature of chemical cohesion. The issue of molecular conformation that seems to represent a classical rather than a quantum-mechanical feature of a molecule will now be revisited with the same objective. This endeavour stems from many frustrating efforts to manipulate the quantum-mechanical Hamiltonian of small molecules into a form that reveals a molecular shape corresponding to the structure, well known to chemists from spectroscopic and crystallographic analyses. The nature of the problem was succinctly summarized by Sutcliffe [193], arguing from the premise that solutions of Schrödinger's equation for nuclei and electrons of specified numbers are known and that one is required to derive from this the three-dimensional structure of the corresponding molecule. The underlying symmetries of the problem, considered to be that of a stationary state, are invariance of the Hamiltonian under uniform translations of all particle variables, rigid rotation-reflections of all particles in any origin, and permutation of variables of identical particles. The spherical symmetry implied by the rotation-reflection invariance would seem to militate against any structure with a geometrical shape. Then there is the problem that, if a given Schrödinger equation describes the molecule of choice, it also describes equally all reactants from which that molecule and all its isomers can be formed and all products into which the isomers can decompose. The chance of finding a unique structure from this solution is clearly nil.

The problem is conventionally sidestepped by assuming that nuclear and electronic motions are decoupled, but despite many efforts this condition has never been shown to yield a rigid molecular shape either. The insurmountable problem is permutational invariance. In molecular-orbital calculations that decouple electronic from nuclear motion the nuclei are identified in order to support the definition of molecular structure, but then permutation of identical nuclei implies rearrangement of bonds and a new set of calculated electronic energies. There is little hope of ever overcoming these problems;

it is therefore often concluded that an isolated molecule has no quantum-mechanically defined molecular shape. The shape it is commonly assumed to have must be a classical concept superimposed on the quantum-mechanical model, and hence nothing more than an assumption *ad hoc*.

The possibility that conformation as an inherent molecular property is revealed within the Bohm interpretation has never been tested. The fact that double bonds show an experimentally measurable barrier to rotation strongly suggests that there is a conformational factor at work and this could serve as a starting point for examination of the broader issue.

6.4.1 Barriers to Rotation

The traditional $\sigma - \pi$ description of multiple bonding and the expectation that conformational effects could somehow be derived from a molecular Hamiltonian function can both safely be discounted. Looking for a fresh approach in terms of quantum potential, returns the argument to the general form of polar wave functions

$$\Psi = Re^{iS/\hbar},$$

recalling that $\nabla S/m$ defines a velocity field. This explains why electrons have zero kinetic energy in stationary states with real wave functions. It is noticed that the only stationary states with complex wave functions occur for the quantum number $M_l \neq 0$. It is significant to note that these are the states that would allow coupling with the magnetic field of a polarized photon, suggesting that all chiral molecules have eigenstates with a non-zero component of angular momentum in projection along z. This assumption opens the possibility of relating conformational effects to a vector quantity, which is more sensible than to expect a scalar function like energy to generate three-dimensional shapes. This important role of angular momentum as a structure-generating factor would normally remain hidden. A good example is provided by the barrier to rotation in the ethylene molecule.

Ethylene C_2H_4 is the simplest molecule to exhibit an electronic (non-steric) barrier to rotation. If this molecule is assumed to contain a C-C linkage, a special direction may occur in one of two possible ways

Of the twelve valence electrons four have non-zero angular momentum ($l = 1$) of $L = \sqrt{2}\hbar$. For two of these, with $m_l = 0$ the z-component $L_z = 0$ and to ensure overall quenching the remaining two electrons should have

$m_l = +1$ and -1 respectively. The charge circulation and distribution for the two possible conformations are shown in figure 5.

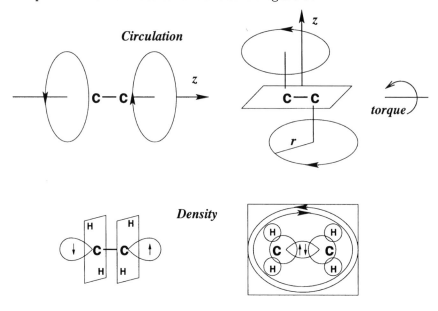

Figure 6.5: *Two possible modes of quenching orbital angular momentum in the ethylene molecule. Only the second possibility leads to a planar molecule with a barrier to rotation and fixed positions for the hydrogen atoms.*

According to figure 3.8 the p-electron density will be concentrated in a lobe along z and two doughnuts concentrated in the xy-planes through the carbon atoms. In the case where z is perpendicular to the C-C bond the two doughnuts overlap in the same plane and much more efficient overlap with the hydrogen $1s$ electrons is possible. A planar structure, eclipsed along the bond direction is predicted. The alternative arrangement would produce a staggered configuration. The staggered arrangement only has a steric barrier to rotation. The angular momentum of the eclipsed conformation however, would resist any torsion of the bond. An applied torque that twists the central bond lowers the symmetry that relates the two vectors $m_l = \pm 1$ and generates a non-zero component of angular momentum.

The generation of angular momentum constitutes the barrier to rotation. In order to torque a centrosymmetric system into a state with non-vanishing angular momentum it is necessary to provide the kinetic energy required to initiate charge circulation. It follows that neither barriers to rotation nor the strenghts of double bonds depend on the overlap of π-orbitals. The more logical conclusion is that barriers to rotation occur whenever an applied torque changes the angular momentum.

Although the electron distribution predicted by the angular-momentum model is essentially the same as that obtained in terms of the conventional scheme of sp^2 hybridization, the interpretation is exactly the opposite. The barrier to rotation is here ascribed to the p_{xy} quenching of angular momentum while the conventional scheme involves the overlap of p_z orbitals.

To calculate the energy barrier to rotation it is noted [35] that the kinetic energy of a rotating charge at a distance r, is

$$T = \frac{m_l^2 \hbar^2}{2mr^2} \tag{6.2}$$

The radius of charge circulation in the twisted molecule is not specified but should be of the order of the C=C bond length. In fact, using $r = 117$ pm, the calculated energy of 270 kJmol^{-1} agrees with the accepted π-bond strength in ethylene.

All of the information that was used in the argument to derive the D_{2h} arrangement of nuclei in ethylene is contained in the molecular wave function and could have been identified directly had it been possible to solve the molecular wave equation. It may therefore be correct to argue [161, 163] that the *ab initio* methods of quantum chemistry can never produce molecular conformation, but not that the concept of molecular shape lies outside the realm of quantum theory. The crucial structure-generating information carried by orbital angular momentum must however, be taken into account. Any quantitative scheme that incorporates, not only the molecular Hamiltonian, but also the complex phase of the wave function, must produce a framework for the definition of three-dimensional molecular shape. The basis sets of *ab initio* theory, invariably constructed as products of radial wave functions and real spherical harmonics [194], take account of orbital shape, but not of angular momentum.

To understand the formation of a triple bond between two CH fragments it is noted that no more than two p-electrons can be directed along the C-C (z) axis, allowing the formation of a linear $H(s)^1C(sp)^1C(sp)^1H(s)^1$ molecule. The remaining pair of p-electrons circulate in the xy-plane with opposite angular momenta. There is no barrier to rotation. The conventional description of a triple bond in terms of one σ and two π interactions is inconsistent, not only for the reasons already discussed, but also because it violates the conservation of total angular momentum when assigning 2 pairs of p-electrons to p_x and p_y orbitals, in addition to the p-density in the sp hybrid orbitals. There is a total of only four p-electrons ($l = 1$) in the system.

The third hydrogen is therefore spread out in a girdle of proton density around the central B atom.

When BH_3 is compared to BF_3 an important difference is that the bonding electrons in the latter case contributed by the ligand atoms are of p-type so that $2 \times s^2$ and $2 \times p^4$ electrons are available in the valence shell of the central atom. This distribution is consistent with the familiar $3 \times (sp)^2$ formulation that defines a planar molecule with a 3-fold axis along z. The angular momentum is quenched by pairs of p-electrons rotating in opposite sense around z.

Apart from the difference in geometry a difference in chemical reactivity between BH_3 and BF_3 is also implied. As in BH_3 each B forms two B-H

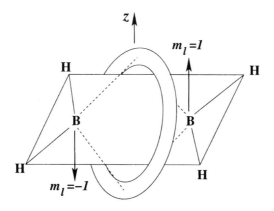

Figure 6.6: *Schematic drawing of the B_2H_6 molecule with angular momentum vectors showing the origin of a barrier to rotation.*

bonds that involve $B(s)$ and $B(p)$ electrons, together with 2 $H(s)$. Combination of these fragments requires anti-parallel alignment of the $p(xy)$ angular momentum vectors, giving rise to a planar B_2H_4 arrangement.

The central electronic region consists of $B(s)H(s)^2B(s)$, consistent with an annular distribution of the two hydrogen nuclei, as shown in figure 6. However, by symmetry the bridging hydrogen nuclei could localize near the z-axis, to reproduce the experimentally known structure.

Hydrides of Carbon

The simple hydrides of carbon include CH, CH_2, $(CH)_2$ and $(CH_2)_2$. Of these, $CH(s^3p_z^2)$ has a non-classical barrel-shaped structure, as shown in figure 7, with hydrogen cylindrically delocalized around C. The bonding electrons, a non-bonding pair and one unpaired electron are distributed in an annulus

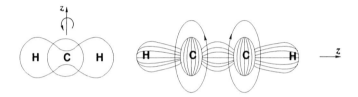

Figure 6.7: *Structures of CH, CH$_2$ and (CH)$_2$ molecules.*

of overlap between H(s) and C(s) regions. Quenching of the orbital angular momentum of two p-electrons in p_z is not an option. CH$_2$ has the same structure as CH, but where the extra H(s) electron now combines with the previously unpaired electron. Acetylene is a linear molecule, also shown in figure 7.

Ethylene has the well-known classical D_{2h} structure with a barrier to rotation. The next in complexity of the simple hydrides is the methyl radical CH$_3$. The obvious (sp^2) planar arrangement can only accommodate six of the seven valence electrons. The electronic configuration of this molecule can therefore not be described in terms of either atomic wave functions or hybrid orbitals. An alternative approach is to view the structure of the methyl radical as a reduced-symmetry form, derived from the structure of methane, to be considered next.

The C valence shell in the methane molecule can be formulated in terms of either $(p_z)^2$ or $(p_{xy})^2$ to be consistent with quenched angular momentum. To derive a molecular shape it is necessary to accept that the four hydrogen atoms are equivalent by molecular symmetry. Any of the four H-C directions may then be defined to coincide with the z-axis of the molecule. The two angular momentum vectors ($l = 1$) cannot quench along this axis since that would violate the equivalence assumption. The only alternative is circulation in two planes ($m_l = \pm1$) perpendicular to z. This description must be valid from the perspective of any of the four hydrogen atoms. Rotation of the two boldly drawn triangles in figure 8 represents the circulation of charge if the z-axis is selected to lie perpendicular to the plane of the paper. Charge accumulation is predicted to occur at the points of intersection of the eight equivalent planes, which together define a regular octahedron, centred at the position of the carbon atom. This condition, as shown in figure 8, is satisfied only if the p-density is localized on the six equivalent sites between pairs of hydrogen atoms. The electron-density distribution in the predicted non-classical structure is radically different from that in the geometrically equivalent classical structure and the tetrahedral symmetry occurs for completely different reasons.

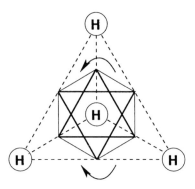

Figure 6.8: *The structure of methane.*

The other major type of covalent bonding is epitomized by the aromatic hydrocarbon benzene, $(CH)_6$. Two possibilities deserve consideration, corresponding to the classical Kekulé and Ladenburg structures with respectively two and three carbon-carbon bonds at each bonding centre. The two possibilities correspond to respective valence shells made up of either 7 or 8 electrons per carbon atom. It is noted immediately that the Ladenburg trigonal prismatic structure does not satisfy the condition of equivalent bonds and must be discounted. The only other arrangement that allows six equivalent carbon atoms is an octahedral one. However, the four covalent bonds required to link each carbon atom to its six neighbours saturate the valence shell and leave no room for the formation of C-H bonds.

Each carbon atom has residual angular momentum which can be quenched in a Kekulé configuration, provided that all atoms are in the xy-plane, as shown in figure 9.

When the same logic and local trigonal symmetry that predict the aromatic arrangement for benzene is applied to $(CH)_8$ a planar structure is ruled out by geometry and hence angular momentum cannot be quenched in a single plane. Alternating double and single bonds are predicted for the non-aromatic cyclo-octatetraene.

There are no steric restrictions that militate against the formation of planar $(CH)_5$. However, by the same principle that stabilizes benzene, $(CH)_5$ should have an unpaired spin and residual angular momentum of $\hbar$ along z, because of the odd number of p_{xy}- electrons. In order to quench the angular momentum one p_z-electron could transfer to p_{xy} and to pair the odd spin, the molecule may accept an extra electron. The most efficient solution would be accepting an electron with angular momentum, *e.g.* by photochemical electron transfer. This may account for the ease whereby $(CH)_5^-$ sandwiches

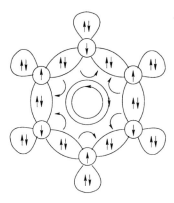

Figure 6.9: *Spin and orbital angular vectors in the benzene molecule.*

transition-metal cations.

Ammonia

Ammonia is one of the few examples of common molecules known to have a non-classical structure [195]. The valence shell consists of 3 p and 5 s-type electrons. Quenching of angular momentum requires $m_l = +1$, 0 and -1

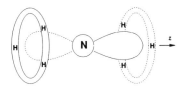

Figure 6.10: *Predicted non-classical structure of the ammonia molecule*

for the three p-electrons. The absolute direction of z is not fixed and the density must be assumed to extend along $\pm z$ with equal probability. This ambiguity carries over to the xy-plane that contains the hydrogen nuclei and predicts the well-known non-classical inverting umbrella structure, stabilized by tunnelling through the nitrogen site.

Water

The oxygen valence shell in H_2O contains 4 p-electrons. Viewed down z, quenched angular momentum requires either $(p_{xy}^2)p_z^2$ or $(p_{xy})^4$ and hence this direction cannot coincide with an O-H bond if the oxygen shell consists of four

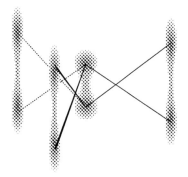

Figure 6.11: *The non-classical inverting umbrella structure of ammonia [195].*

equivalent pairs. The second option defines no special direction and produces no structure at all. The first option implies a p_z lone pair and a two-fold axis

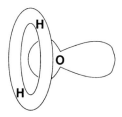

Figure 6.12: *The structure of water.*

through oxygen in the plane containing the three nuclei, consistent with the configuration

$$\{H(s)^2O(p_{xy})^2\} \cdot \{O(p_z)^2O(s)^2\}$$

The hydrogen nuclei are distributed over a torus in a plane perpendicular to z and the resulting non-classical structure is indistinguishable from the recognized classical structure [196] with lone-pair density along the external H-O-H bisector.

Hydrogen Fluoride

The electronic structure of HF in terms of the fluorine valence shell with 5 electrons and z along the internuclear axis, consists of

$$\{F(p_{xy})^4F(s)^2\} \cdot \{F(p_z)H(s)\}$$

This conformation is identical to a classical H-F bond and justifies the p assignment of F electrons in BF_3 above.

Diatomic Molecules

Many known facts about the nature and strength of the bonds in first-row homonuclear diatomic molecules can be accounted for in terms of the simple rule that two p_{xy}-electrons with common m_l constitute an inert pair that resists mixing with other states. Singly occupied p_{xy}-states, on the other hand, readily mix with an s-state of comparable energy. The rationale behind this tendency is that the radius of charge circulation contracts while spreading the s-density, and so facilitates the formation of a chemical bond. The mixing of s and p_z states only happens when no p_{xy}-state is occupied. Application of these rules to first-row diatomics is summarized in table 1 that shows the configuration of atomic valence states, details of the nuclear screening and bond orders. The z-direction coincides in all cases with the bond and is an axis of total rotational symmetry. Narrow vertical ellipses represent degenerate p_{xy}-states with valence-state charge circulation shown where appropriate. C_2 and O_2 are predicted to have triplet ground states and bond order of two as discussed below. The predicted bond orders of N_2, B_2 and F_2 are respectively 3, 1 and 1.

6.4.4 The Role of Screening

The empirical finding that increased strength of multiple bonds over electron-pair bonds is mainly caused by an increase in single-bond strength at the closer interatomic approach that becomes possible due to screening of internuclear repulsion, can now be examined more closely. Whenever the number of valence-electron pairs on an atom exceeds the number of electron-pair bonds to that atom, the excess density may screen the nucleus. Screening becomes effective when the excess density occurs in atomic s-type states with an appreciable s contribution. The first-period diatomics considered before illustrate this screening condition well.

In the C_2 and O_2 molecules with $s^2 p_z^1$ bonds, the excess density in sp_z, but not that in p_{xy}, screens the nuclei, though less effectively than s-pairs. For N_2 there are two s-pairs and this raises the bond order to three. Since inert p_{xy} pairs have no screening effect, F_2 has bond order unity.

The increased strength of the central bond in ethene is best understood in terms of a simple molecular, rather than an atomic interaction. The central bond is established in the xy-plane by interaction between two CH_2 fragments in the valence state $\{C(sp_z)^1 \cdot 2 \times H(s)^2 \cdot C(p_{xy})^1\}$, as shown before in figure

Table 6.1

Valence state	Bond diagram	Screening	Bond order
$B(sp_z)^3$		None	1
$C(sp_{xy})^3 p_z^1$		sp	2
$N\, s^2 p_{xy}^2 p^1$		s^2	3
$O(sp_{xy}^+)^3 (p_{xy}^-)^2 p_z^1$		sp	2
$F(sp_z)^3 (p_{xy})^4$		None	1

Table 6.1: *Details of homonuclear first-period diatomic molecular structures.*

5. It relates to the 50 percent s-character of the sp_z linear combination that allows the excess density to screen the internuclear repulsion. Acetylene $[2 \times \{H(s)C(p_z)^2\}s^2]$, like N_2 has two excess pairs in s-states to screen the nuclei and generate the bond order 3.

As a test these rules are applied to determine the bond order of the carbon-oxygen bonds in CO_2 and CO. The configuration of CO_2 is written as

$$p_{xy}^2 (OCO)^{12} p_{xy}^2$$

where $(OCO)^{12} \equiv \{O(sp_{xy})^3 (p_z)^1 \cdot C(p_z)^1 s^2 (p_z)^1 \cdot O(p_z)^1 (sp_{xy})^3\}$. This predicted arrangement matches the common O=C=O formulation according to the screening formula of table 1. Bond order 2 arises from (sp) screening. The simplest formulation of CO is in terms of $C(s^2 p_{xy}^1 p_z^1) - O(p_z^1 p_{xy}^3 s^2)$, as in N_2, corresponding to bond order 3.

The relationship between screening and bond order is obvious. Bond order 3 relates to the presence of two s^2 lone pairs. In benzene, neighbouring nuclei are screened by the equivalent of one sp electron. The predicted bond order is fractional and $\simeq 1.5$. In general integral bond order is the exception rather than the rule.

6.4.5 Molecular Geometry

The key to the understanding of molecular geometry is in the definition of an holistic molecule described by one of the factors in a product function, equation 5.7. This factor represents the molecular wave function for the molecule of interest and defines non-local entanglement within a limited region of space that varies as a function of the environment. The relative confinement of the molecule means that the boundary conditions on the molecular wave function are variable, in the same sense as for the compressed atom. Different solutions are therefore found in different settings and particularly in different states of aggregation. Any structure revealed by the wave functions must likewise be a function of the surroundings.

The truly isolated molecules of Nature in intergalactic space are rather inaccessible for study. They include the binary hydrides of the light elements mostly with non-classical structures. It is safe to assume that, apart from shape, these molecules will have comparatively large molecular volumes and relaxed cohesion. The bonding forces will be the most elementary, consistent with a stationary ground state and the conservation of angular momentum. The interplay of these factors will automatically be optimal because of non-local interaction within the molecule and fully expressed in the solutions of the wave equation under the appropriate boundary conditions. However, the only partial solution ever obtained was for the dihydrogen ion H_2^+, with the nuclear distribution treated classically, and the chances of a full quantum-mechanical solution for any molecule are slim. It is nevertheless possible to improve in principle on traditional quantum-chemical calculations by starting from theoretically more reliable models of nuclear arrangement and the predicted gross electron distribution. An improved optimization should likewise be possible by including the the angular momentum explicitly in the calculations.

To continue on the theme of calculating molecular conformation it is noted that any non-planar four-atom fragment presents the problem of discrimination between torsional forms of opposite sign. The two forms could be energetically equivalent in all respects, but still differ in an absolute sense, being of opposite handedness, *i.e.* having enantiomeric chiral structures. If the electronic distribution for the fragment has residual angular momentum $\mathbf{L_z}$, the z-direction discriminates between the enantiomers in an absolute sense;

the molecular quantum number M_l have opposite signs in the two cases. Although the molecular wave equation cannot be solved and the eigenvalues L_z not determined *ab initio*, vector addition of atomic angular momenta is possible in principle, and in this sense molecular shape is projected from the wave function by an angular-momentum operator.

6.4.6   Nature of the Chemical Bond

Having examined the stabilization of several different molecules, precious little seems to be left of the cherished concept of electron-pair bonding. It is only in a few special cases such as the familiar aliphatic single bond that the notion of a localized bonding pair can be argued with any confidence. The calculation of bond energies in terms of electron pairs [135, 142] nowhere implies that the bonding pair accumulates in the internuclear region. It is only in the abstraction of isolating a pair of neighbouring atoms for study, that the idea of an electron pair in the Lewis sense could be implied. The distribution of valence density in a molecule is dictated by quenching of orbital angular momentum and could end up at sites commonly identified as bonding regions. Only in these special cases would the classical description of a covalent bond be correct, by accident.

The more general view is that non-local interaction within a molecule determines the charge distribution and conformation holistically. All local features are consequences of the whole. However, molecules of any complexity are rarely the product of a one-step reaction starting from the atomic constituents and are more likely built up from intermediate fragments that retain some of their own molecular properties on incorporation into a bigger whole. This mechanism explains the large number of additive rules that have been discovered empirically for molecular systems [51] and the existence of isomers. A molecule whose conformation and properties are functions of its chemical history, is not holistic, but partially holistic [2], which means that its wave function is a product function, albeit with a limited number of factors (fragments),

$$\psi = \prod_{i=1}^{n} \phi_i$$

where $1 < n < a$, the number of atoms in the molecule. The non-local holistic interaction is a function of fragment size (*i.e.* the quantum potential). When a fragment loses its integrity under the influence of holistic interaction intramolecular rearrangement occurs.

6.4.7 Optical Activity

The relationship between orbital angular momentum and molecular chirality is conveniently introduced by reference to the methane structure which is stabilized by the anti-parallel alignment of two angular momentum vectors, figure 8. The balance is exactly symmetrical in point group $T_d : \bar{4}3m$ only if all ligands around the central atom are identical. Although both vectors may change in magnitude, they stay in balance if one of the ligands (along z say) is different from the others, but now with molecular symmetry $C_{3v} : 3m$. The change in magnitude is caused by the modified distance between ligand atoms and hence the positions of density accumulation. Replacing a second H atom (at the top in the diagram say) by yet another atom may change the direction of the vectors, but both are once more affected in the same way. The symmetry at this stage is $C_s : m$ with a vertical mirror plane that contains the two unique ligand atoms. The vectors become disaligned only when this last element of mirror symmetry disappears and molecular symmetry reduces to $C_1 : 1$. At this stage angular momentum is no longer quenched, $(\mathbf{L}_z \neq 0)$, the molecular quantum number M_l is non-zero and polarized photons interact with the resulting magnetic moment. The plane of polarization is affected differently by enantiomers with respective positive and negative values of M_l. The two enantiomers have identical molecular Hamiltonians and energies - they only differ in angular-momentum eigenstates. Decoupling of angular momentum vectors happens whenever a chiral centre, here defined in terms of four dissimilar substituents in tetrahedral relationship, occurs in a molecule.

There is considerable scope for further work on the topic of molecular chirality. There is a clear definition of how to relate optical rotation to degree of chirality in terms of molecular symmetry. The ultimate aim is a theoretical framework for the quantitative interpretation of optical rotary dispersion and circular dichroism of chiral materials. The importance of this pursuit is almost self-evident.

Experimental testing of these ideas could be done by the study of paramagnetic susceptibilities of chiral material. It is a common assumption[197] that orbital angular momentum is completely quenched and that paramagnetism is entirely due to spin. It is not uncommon, however, to find that incompletely quenched orbital angular momentum is invoked to explain experimental deviations from spin-only values. It is inferred that standard instrumentation is sufficiently sensitive to register the magnetic moments here predicted to be associated with chiral molecules. Other techniques that may provide an insight are high-resolution electron paramagnetic resonance and diffraction analyses with ultrahigh intensity neutrons. Such experiments may reveal a suspected, but still undetected asymmetry between chemically equivalent enantiomers. In only one of the forms can the predicted magnetic moment line up with the chirality of the space that dictates the right-hand rule.

6.5   Conformational Analysis

Several concepts such as molecular shape, configuration and conformation can easily be confused. In the following discussion configuration will be equated to connectivity and conformation with the three-dimensional arrangement of atomic nuclei. Molecular shape embraces both other concepts and also defines the external contours of the electron density of a molecule. It will be argued that molecular shape, and by implication both other structure descriptors, can be understood as an image of local space-time structure. The concept of immediate interest, molecular conformation is fully specified by the three-dimensional cartesian coordinates of all atoms that constitute a molecule. An equivalent description, preferred by stereochemists, is possible in terms of the internal coordinates of bond length, valence angle and dihedral or torsion angle. The advantage of this description is that it provides a clear specification of the local geometry of important functional centres, such as an active site. In many cases the conformation of cyclic fragments in a molecule have a similar role.

Problems associated with the quantum-mechanical definition of molecular shape do not diminish the importance of molecular conformation as a chemically meaningful concept. To find the balanced perspective it is necessary to know that the same wave function that describes an isolated molecule, also describes the chemically equivalent molecule, closely confined. The distinction arises from different sets of boundary conditions. The spherically symmetrical solutions of the free molecule are no longer physically acceptable solutions for the confined molecule.

The simplest illustration of this argument is provided by a free particle in linear motion, correctly described by a wave function (T6.2.1) that satisfies the equation

$$H\psi = -\frac{h^2}{8\pi^2 m}\frac{d^2\psi}{d\psi^2} = E\psi$$

i.e.

$$\psi = a\sin kx + b\cos kx \quad , \quad E = \left(\frac{k}{\pi}\right)^2\frac{h^2}{8m}$$

Since there are no restrictions on the value of k the energy spectrum is continuous. When the particle is confined to the line segment $0 \le x \le L$, new boundary conditions, $\psi(0) = \psi(L) = 0$ come into play and these require that only those solutions described by $b = 0$, $a = \sqrt{2/L}$, and correspond to the quantized energy levels

$$E_n = \left(\frac{n}{L}\right)^2 \frac{h^2}{8m}$$

remain physically acceptable. These are not new solutions, only special ones, singled out by the boundary conditions, such that $k = n\pi/L$. The effect of confinement is to impose additional structure on the motion of the particle.

A chemical molecule, by contrast consists of many particles. In the most general case N independent constituent electrons and nuclei generate a molecular Hamiltonian as the sum over N kinetic energy operators. The common wave function encodes all information pertaining to the system. In order to constitute a molecule in any but a formal sense it is necessary for the set of particles to stay confined to a common region of space-time. The effect is the same as on the single confined particle. Their behaviour becomes more structured and interactions between individual particles occur. Each interaction generates a Coulombic term in the molecular Hamiltonian. The effect of these terms are the same as of potential barriers and wells that modify the boundary conditions. The wave function stays the same, only some specific solutions become disallowed by the boundary conditions imposed by the environment.

The ultimate implication should be clear. A given set of sub-atomic particles may be conditioned in an infinite variety of ways by different environments and histories. All of these constraints should be stipulated in the molecular Hamiltonian in order to avoid confusion with irrelevant permutations. The degree of structure becomes a function of the complete Hamiltonian.

All familiar molecular structures have been identified in the crystalline state. To describe such molecules quantum-mechanically requires specification of a crystal Hamiltonian. This procedure is never attempted in practice. Instead, history is taken for granted by assuming a specific connectivity among nuclei and the crystal environment is assumed to generate well-defined conformational features characteristic of all molecules. Although these decisions may not always be taken consciously, the conventional approach knows no other route from wave equation to molecular conformation.

Despite the vague definition of molecular shape, this concept remains central in chemical thinking [198]. It features at all explanatory levels, shaping important concepts such as chemical reactivity, reaction mechanism, phase transition and reaction pathway. It enters these arguments in the guise of concepts like steric congestion, ligand-field stabilization, barriers to conversion, overcrowding, packing considerations, cone angles, Jahn-Teller distortion, trans effects, steric control, vibronic coupling, transition state, reaction intermediate and many other terms that make a brief appearance in the literature, to explain some poorly characterized observations. This diversity of concepts have one feature in common. They cannot be reduced to fundamental electronic factors and represent some manifestation of molecular conformation. It becomes essential to identify and distinguish between the molecular conformational classes at the root of chemical structure-function relationships.

A first look at molecules differentiates between configurations that are topologically one, two or three dimensional, representing classes of string-like, ring-like and thing-like structures. The most useful parameter to describe conformational features in all these cases is the torsion angle. According to the definition of Klyne and Prelogg [199] the torsion angle about a bond jk in the fragment ijkl is the angle between the projections of the bonds ij and kl on to a plane, perpendicular to the bond jk when viewed in the direction from j to k. The sign of the angle does not change with the direction in which the bond is viewed. The sign convention and the effects of symmetry are illustrated in figure 13. Bonds related by rotation axes have identical torsion angles and

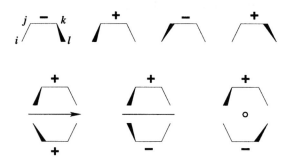

Figure 6.13: *The torsion angle sign convention and the effect of rotation, reflection and inversion operations on the sign.*

those related by inversion axes have equal torsion angles of opposite sign. The orientation of any bond with respect to those adjacent to it is usually obvious from the sign of its torsion angle. Considered in clockwise sense a positive sign indicates an increased elevation, as demonstrated in the sketch below. The torsion angles in a cyclic system of known geometry can therefore readily be specified in terms of these simple principles. Conversely, the symmetry or conformational type of a ring system can be determined qualitatively from a set of experimentally measurable endocyclic torsion angles. However, the actual conformations of cyclic fragments only rarely correspond to exact symmetry types and a more quantitative approach to identify intermediate forms is required.

6.5.1 The Conformation of Cyclic Fragments

The conformational analysis of cyclic molecular fragments is based on the original idea of Kilpatrick, Pitzer and Spitzer [200], as subsequently developed by various workers [201, 202, 203, 204, 205, 206, 207, 208].

Cyclic fragments are common components of molecules and a quantitative description of their conformation is of obvious importance when analyzing the relationship between molecular topology and function. Chemical factors may require rings to be flat, but more often cyclic fragments are non-planar or puckered and can adopt a variety of conformations.

The smallest possible ring is three-membered and simple geometry demands it to be flat, hence only one conformation is possible. Triangular fragments need not have D_{3h} symmetry however, since the bond lengths need not be equal. This observation illustrates the essential difference between ring conformation, or shape and ring geometry. The conformation of larger rings will be interpreted in this sense, *i.e.* as independent of differences in geometry. A four-membered ring can be puckered in only one sense, with alternate positive and negative displacements of atoms from the mean molecular plane. Ignoring differences in geometry, this mode of puckering can always be represented by equal displacements of the corners of a square:

The diagram illustrates a general mode of puckering that occurs for all even-membered rings. The resulting conformation is known as the *crown* form and derives from a regular distortion of the planar D_{Nh} polygon.

The five-membered ring is the simplest odd-membered ring with pucker. Two symmetrical modes, based on the regular pentagon are represented by:

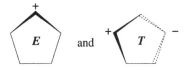

The former has a σ_v mirror plane and the latter a two-fold axis in the mean plane. For distinguishable atoms five equivalent forms of each type are distinguished by the position of the symmetry elements. Each of these has an enantiomeric form, yielding a total of twenty different puckered forms in two groups of ten conformations with equivalent shapes. For a general N-membered ring this implies a maximum of $4N$ related conformations in equal groups with C_s and C_2 symmetries respectively.

Detailed comparison of puckered forms should consider not only the mode of puckering, but also the amplitude. A general expression that combines mode and amplitude for any ring, derives from the formula for out-of-plane displacements of a continuous ring [209], at the regular positions, θ:

$$z(\theta) = \rho_m \cos(m\theta + \phi_m)$$

where ρ_m is the amplitude and ϕ_m the phase for the mth mode. Noting that

$$z(\theta) = \rho_m(\cos m\theta \cos \phi - \sin m\theta \sin \phi)$$

it follows that the displacement at any point is a linear sum of a cos and a sin mode, *i.e,*

$$\begin{aligned} z(\theta) &= \rho_m \cos \phi \cdot \cos m\theta - \rho_m \sin \phi \cdot \sin m\theta \\ &= A \cos m\theta + B \sin m\theta \end{aligned}$$

These modes are $\pi/2$ out of phase (ϕ) and from the graphical representation

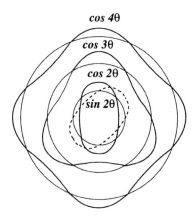

Figure 6.14: *The displacement modes of a circular rod. The trigonometric functions are plotted using reference circles as zero amplitude. Positive displacements are on the outside of the reference circles.*

of these modes, shown in figure 14, the cos mode at $\phi = 0$ and $m = 2$ is seen to correspond with the E form of a five-membered ring, while a T form occurs at $\phi = \pi/2$. The E conformations occur at $\phi = k\pi/N$ and the T conformations at $\phi = (k + \frac{1}{2})\pi/N$, $k = 0, 2N(10)$. The symmetrical E and T forms therefore occur alternately at regular phase intervals of $k\pi/2N$, as shown in figure 15. When the barrier to conversion between neighbouring

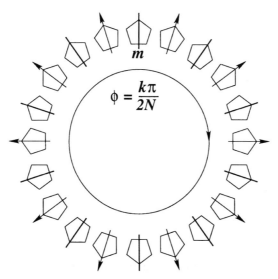

Figure 6.15: *Different forms C_s and C_2 of a five-membered ring placed along a unit circle at phase angles defined by $\phi = k\pi/2N$, $1 \le k \le 20$.*

between neighbouring C_2 and C_s forms is low, the sequence of conformations in order of phase angle constitutes a pseudo-rotational cycle.

The conformation of any of these puckered forms is described by specifying both amplitude and phase and the precise description of an intermediate form can therefore not always be derived by interpolation, using a graphical description such as figure 15, since amplitudes are ignored in this process.

Quantitative specification of intermediate forms is only possible in terms of a linear combination of primitive forms that correspond the group-theoretic representations of normal displacement modes. For regular N-gons the expression that represents normal displacement of the jth atom reduces to [207]:

$$
\begin{aligned}
z_j &= \sum_m \{\rho_m \cos\phi_m \cos[2\pi(j-1)m/N] - \rho_m \sin[2\pi(j-1)m/N]\} \\
&= \sum_m \{b_m \cos[2\pi(j-1)m/N] - c_m \sin[2\pi[2\pi(j-1)m/N]\}
\end{aligned}
$$

For each m the z_j represent an E_m displacement mode of the D_{Nh} symmetry group. It can be shown [203] that, in general, any puckered configuration of an odd-membered ring is represented by

$$\Gamma(\text{odd}) = \sum_m E_m'' \quad , \quad 2 \leq m \leq (N/2) - 1$$

For an even-membered ring the crown displacements are the basis of a one-dimensional representation, B_2, hence

$$\Gamma(\text{even}) = B_2 + \sum_m E_m \quad , \quad 2 \leq m \leq (N/2) - 1$$

The puckering parameters of a general ring are now expressed as a linear combination of $N - 3$ primitive forms, *i.e.* $(N - 3)/2$ pairs of cos and sin forms for all E_m'' of odd-membered rings and the $(N/2) - 2$ E_m pairs, plus B_2 for even-membered rings [208]. For normalized displacements

$$z_j = \sqrt{\frac{2}{N}} \left[\frac{q(-1)^{j-1}}{\sqrt{2}} + \sum_m \rho_m \cos \phi_m \cos \left\{ (j-1)(2\pi m/N) \right\} \right.$$
$$\left. - \sum_m \rho_m \sin \phi_m \sin \left\{ (j-1)(2\pi m/N) \right\} \right]$$
$$\text{defined as} \quad = \quad a(\text{crown}) + \sum_m b_m(\text{cos}m) + \sum_m c_m(\text{sin}m)$$

where $a \equiv 0$ for odd-membered rings, $b_m = \rho_m \cos \phi_m$ *etc.*

The $N - 3$ primitive forms are linearly independent and independent of phase. A complete description of ring conformation is therefore possible in terms of normalized linear coefficients and the integers, k that identify the phases of the contributing forms as $\phi_m = k\pi/2N$, *i.e.*

$$\chi = a(1) + E_m \left\{ b_m(k_m) + c_m(k_m \pm 1) \right\}$$

The correlation of the primitive cos and sin forms with established nomenclature for medium-sized rings is summarized in table 2. In practice the procedure is to start from atomic coordinates and to generate puckering displacements, which are converted to puckering amplitudes and phases, and finally decomposed into a linear sum of primitive contributions:

$$(xyz) \rightarrow z_j \rightarrow (q, \rho_m, \phi_m) \rightarrow a(1), b_m(k_m), c_m(k_m \pm 1)$$

N	$\cos(m)$		$\sin(m)$	
	2	3	2	3
5	envelope		twist	
6	boat		twist	
7	boat	chair[a]	twist-boat	twist-chair[a]
8	boat-boat	twist-chair	boat-boat	twist-chair
9[b]	boat-boat	boat-chair	twist-boat-boat	twist-boat-chair

Table 6.2: *Relationship between the primitive forms and some traditional classical conformations.* [a]The E_3'' representations for seven-membered rings do not correspond with low energy C/TC forms, but have a chair/twist-chair shape. [b]The E_4'' representations for nine-membered rings do not correspond with any established forms.

The first step consists of finding the mean ring plane that defines the displacements z_j. The second step is to solve from the set

$$\rho_m \cos\phi_m = \sqrt{2/N} \sum_j z_j \cos[(j-1)(2\pi m/N)]$$

$$\rho_m \sin\phi_m = -\sqrt{2/N} \sum_j z_j \sin[(j-1)(2\pi m/N)]$$

for all (ρ_m, ϕ_m), and to calculate

$$q = \sqrt{1/N} \sum_j (-1)^{(j-1)} z_j$$

These are the well-known puckering parameters first defined by Cremer and Pople [205].

Using the calculated ϕ_m, the integers k_m for the appropriate cos and sin forms to be used in the linear expansion are determined next. Since the z_j displacements can be expressed either in terms of q, ρ_m and ϕ_m, or as the linear sum $a(1) + b_m(\cos m) + c_m(\sin m)$, the coefficients have been shown [208] to follow as

$$a = q$$

$$b_m = \frac{\rho_m \sin[\phi_m - (k_m \pm l)(\pi/2N)]}{\sin(\mp l\pi/2N)}$$

$$c_m = \frac{\rho_m \sin[(k\pi/2N) - \phi_m]}{\sin(\mp l\pi/2N)}$$

l is an integer and depends on the symmetry of the primitive forms. The final step is to normalize the coefficients.

6.5.2 Molecular Shape Descriptors

To explore the relevance of symmetry allowed conformations to chemically real structures it is necessary to locate steric-energy minima in a potential energy surface that spans all possible conformations. A variety of computational techniques [210] are available, commonly combined with experimental results retrieved from structural databases [211]. Such procedures have revealed the occurrence of countless different rotamers and conformers, arising from pseudorotation and conformational inversion under special environmental conditions. In addition, situations of disorder in the crystalline state are symptomatic in many cases, of the stabilization of variable intermediate forms.

On this basis ring conformation is interpreted to depend on the environment, rather than on chemical bonding. The most regular predicted conformations occur only by way of rare exception, even for cyclic alkanes. Different conformations are routinely observed in different states of aggregation. In the gas phase the entire Boltzmann distribution of conformations is present at any given temperature. Molecular shape and conformation are therefore undefined in this instance.

At another level molecular shape is linked to the external surface of a molecule. Although it is generally recognized that quantum-mechanically molecules do not have clearly defined surfaces, new definitions of molecular shape and surface appear in the literature on a regular basis. Variables such as molecular surface area and volume are useful in the analysis of molecular recognition and other surface-dependent properties that assume a clearly defined surface.

One of the conceptually simplest representations of a molecule, also used in the construction of hard-sphere space filling models, is obtained by centring spheres of suitable Van der Waals radii at the positions of the atomic nuclei. To convey a feeling of size the aggregate is enclosed in a generating ellipsoid that circumscribes the atom most remote from the centre to define the volume of the Van der Waals body [212].

A more realistic outline of a molecular surface can be defined in terms of the outer contours of electron density according to Bader [173]. To avoid excessive computation the densities of large molecules may be built up from previously calculated densities of smaller fragments [213]. The most obvious approach, to approximate molecular density by the sum of atomic densities over the promolecule has also been explored [214]. This approach works well since the deformation density associated with bond formation is small compared to the total density [215]. The total density may therefore be represented by a sum over spherically averaged atomic densities, $\rho(\boldsymbol{r}) = \sum \rho_A(\boldsymbol{r})$.

Using atomic densities calculated from tabulated atomic wave functions, the summation was found [214] to produce results equivalent to the most elaborate molecular Hartree-Fock calculations for a series of small molecules, at a fraction of the computing expense. Surface areas and volumes computed by the two methods were found virtually identical. The promolecule calculation therefore has an obvious advantage in the exploration of surface electron densities, surface areas and molecular volumes of macromolecules for the analysis of molecular recognition.

These results are hardly surprising. A sober re-examination of *ab initio* methods of calculating molecular densities (T 7.4.5) confirms that all structure optimizations start from an informed guess of a trial geometry. The basis sets used to synthesize molecular wave functions are the same as those used in another context to calculate individual atomic wave functions and densities. Subsequent computation of an endless number of integrals contributes no more to the final result than an estimate of the charges exchanged during bond formation. As already stated, these are minor effects in comparison with the total density [215]. The ultimate reason is that molecular cohesion arises from the molecular quantum potential and not from a macroscopic redistribution of charge.

The atoms-in-molecules partitioning of electron density (6.3.1) can now be seen in different perspective. The total crystallographically measured electron density is essentially that of the promolecule, which by definition must partition into atomic densities. Calculated densities, on the other hand, can only be obtained after assuming a set of nuclear coordinates. Partitioning into a set of atomic basins therefore simply demonstrates a degree of self-consistency between synthesis and analysis of the density function.

The use of molecular shape in the study of molecular recognition has also been shown to produce useful results without complicated *ab initio* calculations [148]. Molecular recognition involving macromolecules depends on complementarity of shape, electrostatic complementarity and the matching of non-polar regions. These factors may not always predict the same orientation of partners for optimal locking and an energy criterion is finally required. The energy gain on aggregate formation may be considered made up of three contributions, ΔE_s, ΔE_d and ΔE_{es}. The gain in skeletal energy is strictly intramolecular. It includes all geometrical rearrangements of the proximal partners and is always positive and small. Gain in dispersal energy is essentially intermolecular, negative and the leading contribution to ΔE. It comprises all interpartner interactions, apart from the electrostatic, which is quite small.

In order to study the effects of aggregation between two partners it is nec-

Chapter 7

The Chemical World

7.1 Introduction

In the course of this work it has been necessary to link the fundamental properties of numbers, photons, electrons, molecules and matter to the geometry of space-time, albeit without further insight. An effort will now be made to specify geometrical details consistent with experimental observation.

The philosophical position that the fundamental laws of physics describe various aspects of the geometry of space-time is widely accepted. The first successful demonstration of such a relationship was achieved by Einstein with the formulation of the general theory of relativity that defines gravity in terms of space-time curvature. In a separate development the work of Weyl and Schrödinger resulted in the definition of quantum phenomena in terms of gauge transformation, as a geometrical effect. As a corollary the electromagnetic (photon) field arises from local gauge invariance (T 4.5).

Weyl's original suggestion was that parallel transplantation of a vector ϕ_m changes its norm by an amount

$$\delta l = (\phi_m \mathrm{d}x_m)l$$

that could be related to the potentials of the electromagnetic field. However, if an atom were carried around a closed path in an electromagnetic field this theory predicted that the atom should radiate at a different wavelength at the end of the loop. This prediction is refuted by experiment. Resolution of the problem originated with Schrödinger's demonstration [219] that a complex phase change rather than a gauge change could be used to deduce the quantum conditions of electronic motion from Weyl's world geometry.

Unification of the gravitational and electromagnetic fields was achieved qualitatively in terms of the Kaluza-Klein (KK) five-dimensional model (4.6).

It was demonstrated by London [220] that the amplitude of De Broglie matter waves, ψ had the same properties as Weyl's norm factor by writing $l = l_o \exp(-2\pi i n)$, as suggested by Schrödinger. In terms of the 5D formulation of Klein he proved that the ratio $\psi/l = \exp(2\pi i\alpha/\hbar)/l_o$ was a constant. It was finally showed by Schrödinger [221] that the complex phase factor constituted the common origin of gravitation and electromagnetism. (Compare 5.9).

7.2 Five-dimensional Space-time

Although only partially successful the KK model clearly demonstrated that the geometry of space-time cannot be adequately described in only 3 or 4 dimensions. Any low-dimensional description must fail and is expected to show up unexplained anomalies that automatically disappear when the analysis is repeated in higher-dimensional space. Space-time singularities such as the big bang may be of this type. The simplest demonstration of such a dimensional effect is seen in curved two-dimensional space, like the surface of a sphere. Within the 2D surface the perceived Euclidean geometry consistently gives anomalous results. One demonstration of the effect is to compare the experimentally measured values of π, found by by two-dimensional observers working in flat, elliptical and hyperbolic space respectively.

The relationship between flat and elliptical surfaces with zero and positive curvatures respectively, is shown in figure 1. A plane intersects the sphere

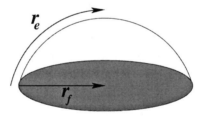

Figure 7.1: *Common circle in elliptical and flat surfaces, respectively.*

in a circle that has the same meaning in the flat and elliptical surfaces, and hence the same circumference. Comparing independent measurements then shows that

$$2\pi_f r_f = 2\pi_e r_e$$

Since $r_e > f_f$, it follows[1] that $\pi_f > \pi_e$. In hyperbolic space, with negative

---

[1]The biblical measurement of $\pi = 3$ (I Kings 7:23; II Chron 4:3) must have been done

curvature, $\pi_h > \pi_f$.

The experimental basis of the theory of special relativity provides another example of such a dimensional effect (T 4.3.3). Within Galilean relativity the line element r^2 is invariant under rotation. The observation that this line element is not Lorentz invariant shows that world space has more dimensions than three. The same effect in 2 and 3 dimensions is demonstrated diagrammatically in figure 2. The norm of the two-dimensional vector is seen to be

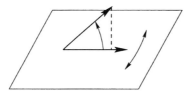

Figure 7.2: *The norm of the vector, invariant to rotation in the Euclidean plane, seems to contract when rotated into a third unobserved dimension.*

invariant under rotation in the plane. On adding a third component to the rotation, the two-dimensional line element (projection) is no longer invariant and a Fitzgerald contraction is inferred.

The KK model has neither been fully accepted nor refuted by the scientific community and it keeps on generating a lot of interest. One of the early criticisms was published by Einstein and Mayer [222], without direct reference to KK. It rejected the KK theory on the grounds that the world is experienced to be four and not five-dimensional; that the assumption of a cylindrically compacted dimension is unnatural; that the theory fails to account for the mass of moving charges; and finally, that the theory offers no physical explanation of the extra coefficient γ_{44}, not associated with either gravitational or electromagnetic fields. To overcome these problems a four-dimensional Riemannian continuum that incorporates a five-dimensional linear vector space was proposed: a five-component vector is defined at each point of the 4D continuum, with fixed rules of combination and transformation. These were chosen to limit the composite space in terms of the coefficient tensors required to define both gravitational and electromagnetic fields, using the same arguments as KK. The proposal of these authors has been forgotten and their objections to KK theory are no longer considered important. In fact, there is growing consensus that world space (T 4.1) is at

---

[2]in elliptical space.

least five dimensional.

The current status of KK theory has been reviewed by Wesson [223]. Kaluza adopted the cylinder condition, setting all partial derivatives with respect to the fifth coordinate equal to zero, and γ_{44} =constant. With another choice of gauge the field equations reduce to:

$$G_{\alpha\beta} = \frac{\kappa^2 \Phi^2}{2} T_{\alpha\beta} - \frac{1}{\Phi} \left(\nabla_\alpha \nabla_\beta \Phi - \gamma_{\alpha\beta} \Box \Phi \right) \tag{7.1}$$

$$\nabla^\alpha F_{\alpha\beta} = -3 \frac{\nabla^\alpha \Phi}{\Phi} F_{\alpha\beta} \tag{7.2}$$

$$\Box \Phi = -\frac{\kappa^2 \Phi^3}{4} F_{\alpha\beta} F^{\alpha\beta} \tag{7.3}$$

where $G_{\alpha\beta}$ (T(4.42)) and $F_{\alpha\beta}$ (T(4.33)) are the usual 4D Einstein and Faraday tensors. $T_{\alpha\beta}$ is the energy-momentum tensor of the electromagnetic field, given by

$$T_{\alpha\beta} = \left(\gamma_{\alpha\beta} F_{\delta\epsilon} F^{\delta\epsilon}/4 - F_\alpha^\delta F_{\beta\delta} \right)/2$$

and $\Box$ is the wave operator.

Equation (2) is recognized as the four equations of electromagnetism modified by a wave-like scalar field. Equation (1) represents the 10 Einstein equations of general relativity, equated to energy and momentum derived from the fifth dimension. In short, KK theory is a unified account of gravity, electromagnetism and a scalar field. Kaluza's case, $\gamma_{44} = -\Phi^2 = -1$, together with the identification

$$\kappa = \left(16\pi G/c^4 \right)^{\frac{1}{2}}$$

leads to

$$G_{\alpha\beta} = \frac{8\pi G}{c^4} T_{\alpha\beta}$$

$$\nabla^\alpha F_{\alpha\beta} = 0$$

the Einstein and Maxwell equations in 4D, here derived from vacuum in 5D.

The KK model contains the standard 4D model, and observations that support the latter also support the former. Applied to cosmological problems an important conclusion is that 5D solutions may reduce to multiple solutions in 4D. Any solution of the field equations depends on a choice of coordinates. It is possible to start with flat 5D space and use coordinate transformations to generate realistic cosmological models such as the standard one. However, the possibility of alternative solutions and explanations pertaining to questions such as the nature of dark matter, the microwave background and the big bang opens up. A corollary is that all of big-bang cosmology could be an artefact produced by an unfortunate choice of coordinates in 5D. The crucial decision in all cases is the definition and interpretation of the fifth coordinate. This issue is still being debated.

7.2.1   The Thierrin Model

An elegant but simple model of a five-dimensional universe has been proposed
by Thierrin [224]. It is of particular interest as a convincing demonstration
of how a curved four-dimensional manifold can be embedded in a Euclidean
five-dimensional space-time in which the perceived anomalies such as coordinate contraction simply disappear. The novel proposal is that the constant
speed of light that defines special relativity has a counterpart for all types
of particle/wave entities, such that the constant speed for each type, in an
appropriate inertial system, are given by the relationship

$$v^2 = nc^2$$

Normal fermionic particles have $n = 1$ and for photons $n = 2$. However,
these velocities are not defined in four dimensions, but in 5D space, made
up of a four-dimensional Euclidean space $S = \{x, y, z, u\}$ and absolute time
$\{t\}$. The first three coordinates of S are familiar in 3D space, with u orthogonal to these in 4D and not observable in 3D. Each (moving or stationary)
particle has its own inertial system S wherein it moves with velocity c along
u. The model therefore contains the surprising results (4.3.2) calculated by
Schrödinger [67] and Winterberg [68] that electrons and photons have intrinsic velocities of c and $\sqrt{2}c$ respectively.

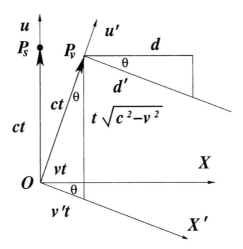

Figure 7.3: *Motion of a particle with velocity v relative to a stationary object*
P_s.

Compare a stationary particle $P_s\{x, y, z, u\}$ and another P_v that moves in
the (x, u) plane with speed v relative to the inertial x-axis of the stationary

particle. The inertial x' and u' axes of $P_v\{x', y', z', u'\}$ are in the same plane as (x, u). The velocity diagram is shown in figure 3. Starting from O at $t = 0$ the particle positions at time t are as shown. From

$$\cos\theta = \frac{v}{v'} = \frac{\sqrt{c^2 - v^2}}{c}$$

the speed of the moving particle relative to X' follows as

$$v' = \frac{v}{\sqrt{1 - v^2/c^2}}$$

Next, instead of a second particle, compare the propagation of a light wave from point O at time 0. After unit time the wavefront has spread to the circle radius $\sqrt{2}c$. At that moment the X axis of the particle's inertial system has moved a distance c along u to intersect the wavefront at $X = c$, as shown in figure 4. Hence the speed of light in this system is measured as equal to

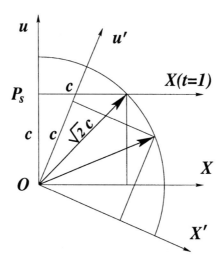

Figure 7.4: *Diagram to illustrate the constancy of the measured speed of light.*

c. The same light wave with respect to the coordinate system of the moving particle clearly yields the same result.

Again from figure 3, the length of two measuring rods d and d' associated with the inertial systems in relative motion, compare as

$$\frac{d}{d'} = \frac{\sqrt{c^2 - v^2}}{c} \qquad \text{i.e.} \qquad d = d'\sqrt{1 - v^2/c^2}$$

The length d' of the rod seen by a stationary observer appears contracted to d. Apparent time dilation arises in the same way.

Now consider the same situation as in figure 3, but with a second particle P_2 rigidly linked to P_v. At $t = 0$ the x-coordinates of P_2 in the inertial systems of P_s and P_v are x_1 and x_1' respectively. The arrangement is shown in figure 5. From simple trigonometry

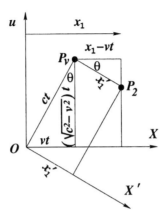

Figure 7.5: *Lorentz transformation between moving frames.*

$$\cos\theta = \frac{x_1 - vt}{x_1'} = \frac{\sqrt{c^2 - v^2}}{c}$$

and hence the Lorentz transformation

$$x_1' = \frac{x_1 - vt}{\sqrt{1 - v^2/c^2}}$$

Finally, suppose that the particle P with kinetic energy $\frac{1}{2}mc^2$ in its inertial system is converted into a photon of speed $\sqrt{2}c$. Its kinetic energy changes to $\frac{1}{2}m(\sqrt{2}c)^2 = mc^2$.

All the important results of special relativity are recovered by simple trigonometry in S_4, treating time as a constant fifth dimension. The sensation of time flow derives from the motion of an observer, stationary in $3D(x, y, z)$, along its u axis.

All stationary objects share a common inertial system and time *appears* to flow in the same direction, u. For moving objects the inertial system is rotated according to the direction of motion, such that time now appears to flow in the new direction u'. Motion, in this context strictly refers to

3D(x, y, z) space only. The 4D motion of both material and energetic entities (photons) is of a different kind. In 3D the former is perceived to be zero and the latter as c. However, this so-called velocity of light is not a velocity in the normal sense of the word.

The standard representation of special relativity in terms of Minkowski space, as in figure 3.2, acquires significant new meaning in the Thierrin scheme. As in figure 6 the light cone now defines allowed directions of the u-axis of material objects and u itself represents the world-line of special relativity. The objection against non-local interaction disappears. Instead, the

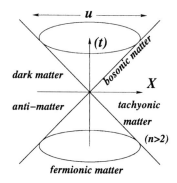

Figure 7.6: *Minkowski diagram of special relativity according to the 5D model of Thierrin.*

existence of different forms of matter may be inferred, including anti-matter and dark matter. There is a striking similarity between the Thierrin model and the unification scheme proposed by Einstein and Mayer [222]. The essential difference is that Thierrin separates time from space coordinates and works in euclidean rather than curved space. The latter difference is minor. It simply means that the u-axes of stationary systems on a curved manifold have different directions, like the time cones of figure 5.13. The former difference probably means that the unification of Einstein and Mayer may be achieved more naturally in Thierrin space-time. There is a similar advantage over KK schemes that add a fifth space dimension to the 4D space-time of general relativity [223].

A significant new feature is that three-dimensional space could be curved into a fourth space dimension without involving the time coordinate. A possible advantage would be that essentially, relativistic effects could be analyzed by the methods of non-relativistic physics - four space coordinates and one time coordinate. It gets around the unpalatable conjecture, already used in this work on several occasions, that time may be flowing in different directions at different points in a curved space-time manifold. In most cases the final conclusions are unaffected, only the model makes more physical sense.

7.3 Topology of World Space

The idea of a closed space-time manifold with an involution has been mooted on the basis of nuclear synthesis (figure 2.6), number theory (figure 2.8), historical argument (4.4), absorber theory (figure 4.8) and chirality (5.9.3). All of these schemes can now be combined into a single construct based on curved Thierrin space-time.

An imperfect lower-dimensional analogue of the envisaged world geometry is the Möbius strip. It is considered imperfect in the sense of being a two-dimensional surface, closed in only one direction when curved into three-dimensional space. To represent a closed system it has to be described as either a one-dimensional surface (*e.g.* following the arrows of figure 7) curved in three, or a two-dimensional surface (projective plane) closed in four di-

Figure 7.7: *Chirality is gradually inverted on transplantation along the surface of a Möbius strip. The same hand is shown on opposite sides of the double cover.*

mensions. A projective plane is obtained by joining diametrically opposed points on the edge of a hemispherical surface, as shown in figure 8.

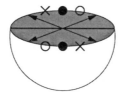

Figure 7.8: *Construction of a projective plane by identification of diametri-cally opposed points of an open hemisphere.*

7.3.1   Space-time Geometry

An obvious problem is that the previous conclusions are not supported by any of the generally accepted cosmological models, which are all based on assumed affine geometries of space-time. Affine geometry differs from other geometries by assuming a unique parallel (through a given point) to a given line [225]. The affine propositions are those which are preserved by parallel projection from one plane to another. These propositions hold, not only in Euclidean geometry but also in the Minkowski geometry of special relativity.

As an example of affine transformation, consider a square transformed under shear and under strain, as shown in figure 9. Lines that originally

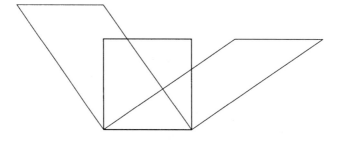

Figure 7.9: *A square, transformed under shear and under strain.*

were parallel remain parallel although angles between lines are not invariant. The resulting figures are equivalent to the original square and to each other. Thus, no distinction is made between squares and parallelograms. Further, no distinction is made between circles and ellipses. The parallel-preserving property of affine transformations means that not all four-sided polygons are equivalent. A square or a parallelogram cannot be transformed into, for example, a trapezium since that would destroy the invariance of parallelism. All triangles are, however, equivalent since no parallel lines are involved.

A more general view is provided by *absolute geometry* that makes no assumption about unique parallels. It therefore includes affine geometry as a special case, but also allows for projective geometry in which there are no parallels, or hyperbolic geometry with more than one parallel through a given point. Elliptic geometry is the non-Euclidean version of affine geometry. It is meaningless to ask which of the geometries is true, and practically impossible to decide which provides a more convenient basis for describing astronomical space. The intuitive guidelines already defined favour a closed chiral space. The first criterion eliminates hyperbolic space and the second rules out affine geometry.

Projective Geometry

Projective geometry was created by Renaissance artists. To make a correct perspective drawing the scene of interest is projected into a plane that intersects all lines between points on the object and the eye of the artist. Affine

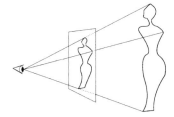

Figure 7.10: *Construction of a perspective drawing from an exterior point.*

properties are no longer invariants in projective geometry. If figures from a given plane are transformed into figures on another, not necessarily parallel plane, by either parallel projection or from an exterior point, as shown in figure 10, the affine invariant of parallelism is lost. However, a straight line remains a straight line and finite configurations remain finite.

As it views a scene the eye does not respond directly to the objects in the scene, but to light rays that travel along straight lines from points in the scene to the eye. It follows that radial lines will look like points and radial planes look like lines. As a result radial dimensions are lost. A further consequence is that non-radial lines acquire an extra point at infinity. As shown in figure 11 the non-radial line L is observed through a set of radial lines in the plane OL. Only one radial line in OL does not connect a point on L to the eye. That one exception is the radial line parallel to L. This line P_∞ appears to the eye as a point at infinity on L. The two parallel lines that meet at infinity are considered to intersect at an angle of zero.

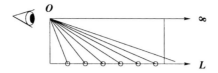

Figure 7.11: *Point at infinity generated in perspective geometry.*

The same reasoning shows that non-radial planes acquire an extra line at infinity. It follows that projective space may be regarded as affine space plus a plane at infinity. One of the most elegant properties of projective geometry is the principle of duality which asserts that, in a projective plane every definition or theorem remains valid on the consistent interchange of the words *line* and *point*.

In the projective plane there is only one kind of conic. The familiar distinction between hyperbola, parabola and ellipse belongs to affine geometry only.

A projective plane may be generated by adding to the Euclidean plane a line at infinity. The Euclidean plane itself is equivalent to the gnomonic projection of a sphere on a plane, σ shown in figure 12. Each point P in the

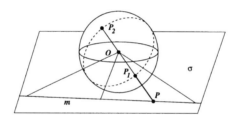

Figure 7.12: *Gnomonic projection of point P_1 on P in the Euclidean plane σ. The line m maps the stippled great circle.*

plane yields a line OP that joins it to the centre of the sphere. The resulting diameter meets the sphere in two antipodal points P_1 and P_2 which are mapped on the same point P. Each straight line m in σ generates a diametral plane that cuts the sphere in a great circle. Conversely each great circle on the sphere, except the equator, which is parallel to the plane, corresponds to a line in σ. This exception can be removed by adding to the Euclidean plane, a line at infinity to represent the equator. Since each line has a single point at infinity antipodal points must be identified and treated as a single point. This operation results in the definition of a projective plane, as shown in figure 8. As before, two parallel lines meet in a point at infinity, and an ordinary line meets the line at infinity in a point at infinity. Hence, any two lines of the projective plane meet in a point.

7.3.2 Topology

In all the geometries considered so far one important invariant under permitted transformation is the preservation of straight lines. Thus, in none of these geometries does a circle, for example, belong to the same equivalence class as a polygon. On proceeding from projective geometry to topology, even this invariant is abandoned. The additional new transformations are referred to as elastic deformations, and include stretching, bending and twisting [226]. Cutting, however, is not permitted and joins may not be made in such a way as to bring together points that were originally separated. Topological equivalence classes include within one and the same class many figures with widely different geometric properties. Topology is particularly powerful in the analysis of continuous surfaces.

An example of topological analysis is shown in figure 13. A rectangular rubber sheet is transformed into a torus by the identification of pairs of opposite sides, first to form an open cylinder which transforms into a torus by joining the two open ends. In technical language, the Euclidean plane is

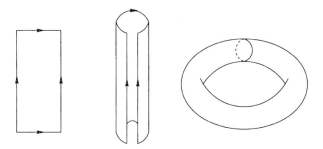

Figure 7.13: *Transformation of a rectangle into o torus.*

the universal covering surface of the torus.

A surface is closed if it has no boundary curves. By this definition surfaces of a sphere and a torus are closed, whilst the surfaces of a hollow cylinder and of a disc are open. Boundary curves of two-sided surfaces are curves which separate one side of the surface from the other, for example the edges of a piece of thin paper. A completely open cylinder has two boundary curves. A cylinder which is half-open has only one boundary curve, and is continuously deformable into, and therefore topologically equivalent to a disc. Similarly, the removal of a disc from the surface of a sphere leaves an open surface with one boundary curve, likewise deformable and topologically equivalent to both a disc and a half-open cylinder. The removal of two separate discs from the surface of a sphere leaves an open surface with two boundary curves, topologically equivalent to a completely open cylinder. The torus is topologically equivalent to a sphere with a handle, like that of a teacup.

So far, all the surfaces considered are two-sided. In order to pass from a point on one side of a given surface to a point on the other side it is necessary to cross a boundary curve. If the surface is closed, then it would be necessary to penetrate the surface in some way. Not all surfaces, however, are two sided. Because of the half-twist required to form a Möbius strip, the resulting surface is one-sided. It is now possible to travel from any point originally on one side of the strip to any point originally on the other side, without crossing a boundary curve. As shown in figure 14 a Möbius strip has only one boundary curve, since another effect of the half-twist is that the

Figure 7.14: *A regular Möbius strip with its single boundary curve. The absolute local curvature of the double cover is constant and the total curvature is zero.*

long edges of the paper, originally opposed, are joined to form a continuous curve, topologically equivalent to a circle. In general, Möbius bands with an odd number of half-twists are one-sided and those with an even number are two-sided [227].

Closely allied to the property of one-sidedness is the property of non-orientability. A surface is said to be orientable if the orientation of an object in the surface is preserved. Consider the handed (chiral) object at a point in the Möbius surface of figure 7. From a local point of view there is a corresponding point on the other side of the surface. Since the Möbius band is one-sided it is possible to draw a continuous path connecting the two points without crossing a boundary curve, as in figure 7. The chirality of the object is reversed when moved along the continuous path between the two points. A situation like this is not possible with two-sided surfaces.

Unlike other closed surfaces the Möbius strip is bounded. The boundary is a simple closed curve, but unlike an opening in the surface of a sphere it cannot be physically shrunk away in three-dimensional space. When the boundary is shrunk away the resulting closed surface is topologically a real projective plane. In other words, the Möbius strip is a real projective plane with a hole cut out of it.

7.3.3 World Space

Matter, in its many forms, occurs with the attributes of mass, charge, spin and chirality. Each of these depends on local space-time structure which, according to Mach's principle, is shaped by non-local interaction [228] through the quantum potential. Fundamental understanding of chemistry, that deals with the transformations of matter, might eventually be possible in terms of the structure and topology of world space (T4.1). Already, there is sufficient evidence to rule out certain notions about the large-scale structure of the universe on chemical grounds:

- In order to support both gravitational and electromagnetic fields, space-time needs at least five dimensions.

- To account for the wavelike nature of matter the universe must be topologically closed to support standing waves.

- The separation of matter and anti-matter can only happen in chiral space-time. This requirement rules out affine geometry.

All evidence points at a multi-dimensional, non-orientable structure, topologically equivalent to projective space-time. It is of interest to note that the same conclusion has been reached before on the basis of astronomical observation [229]. In two instances has the same pattern, defined by a cluster of quasars, been observed as distorted multiple images at different positions in the sky and interpreted in terms of multiply connected projective space.

The only factor of some chemical importance that features prominently in the theories of cosmology is the synthesis of small nuclides such as deuterium and helium. Unfortunately, the initial conditions that are considered to be crucial in these models are purely conjectural. There is little hope of a meaningful test against chemical reality and, in the present climate, no chance for the growth of a mathematically based alternative cosmology. However, the simple qualitative model of a non-orientable universe provides interesting insight into the nature of matter, non-local interaction and quantum theory.

Central to the discussion of space-time structure is the notion of a vacuum. As argued before (5.9) the physical vacuum is not a void. Because

matter exists, space-time cannot be flat (Euclidean) and its curvature produces measurable effects, which can be interpreted as fields that exist in the
vacuum. Even if these fields have no more substance than mathematical formulation, the carriers of a field, such as photons, seem to require a responsive
medium for their transmission. This medium is known as the aether.

The aether concept disappeared from physics for several decades after
the discovery of special relativity, which led to the four-dimensional formulation of all natural laws. The argument against the aether is based on the
perceived nature of a perfect vacuum that contains no matter and no fields.
Relativistically, such a region must be isotropic and all directions within the
light cone must be equivalent. If the aether hypothesis means that a mobile
aether exists at each point in the region, its motion creates a preferred direction within the light-cone in space-time, at variance with the requirement
of relativity. Experiments, like that of Michelson and Morley (T4.3.1), designed to measure the aether drift, actually measure the velocity of the earth
relative to a stationary aether.

It was pointed out by Dirac [230] that the contradiction between relativity
and the aether is resolved within quantum theory, since the velocity of a
quantum aether becomes subject to uncertainty relations. For a particular
state at a certain point in space-time, the velocity is no longer well defined,
but follows a probability distribution. A perfect vacuum state, in accordance
with special relativity, could then have a wave function that equalizes the
velocity of the aether in all directions. The passage from classical to quantum
theory affects the interpretation of symmetry relationships. As an example,
the $1s$ state of the hydrogen atom is centrosymmetric only in quantum, but
not in classical theory. A related redefinition of quantum symmetry provides
the means of reconciling the disturbance of Lorentz symmetry in space-time,
produced by the existence of an aether with the principle of relativity.

Since the symmetrical quantum state of the aether, proposed by Dirac,
is not normalizable it should be considered a theoretical idealization, which
can never be actually realized, although it can be approached indefinitely
closely. Another such a state describes a particle with specified momentum.
In this case the wave function of the particle cannot be normalized because
of the uncertainty principle. By analogy, an aether that conforms to both
quantum mechanics and relativity has to be an unattainable idealized state,
like the perfect vacuum, or void. The real vacuum is an approximation to
and more complicated than that. It is of lower symmetry and structured in
such a way that particle velocities can be specified relative to the aether. To
make any sense, the details of electronic motion through space, demand the
presence of a structured aether.

7.4 The Geometry of Quantum Events

An alternative model of the universe has been proposed, under this same heading, before [28]. The physical vacuum is assumed to contain an element of PCT symmetry in four-dimensional space. Thereby the vacuum is defined as an interface, either between two universes or between two regions of opposite chirality in the same universe. The latter more economical situation is the more attractive. The perfect interface, like the ideal aether does not exist and the experimentally measured properties of the vacuum represent the faint echo of another enantiomeric world from across the interface.

Progressively smaller particles experience, to an increasing extent, the effects of interaction with the hidden world beyond the interface. An observer that keeps track of the particle, unaware of the hidden interaction finds that the motion becomes inexplicably more erratic. The differential equation to model the motion is found to resemble a wave packet rather than a classical particle. The mathematical description of the particle's progress is precise, but the physical interpretation is incomplete. The crucial result is that the particle in the vacuum does not behave differently from classical particles. Its progress depends on the same logic and causality, but since its equations of motion are formulated with neglect of a vital segment of its total environment, they appear more complicated than necessary. This anomalous behaviour decreases rapidly with increasing aggregation. Any surface chemist is familiar with this condition and knows that a sharp potential gradient occurs across an interface.

The formation of an interface in a two-component homogeneous fluid happens when interaction between like entities becomes dominant. The resulting rearrangement is an example of symmetry breakdown that leads to the segregation of components into separate layers. On the cosmic scale a phase separation between matter and anti-matter is assumed to create two three-dimensional worlds in 4D space, such as S_4 of Thierrin space. The interface can therefore not be crossed in three dimensions.

A precise analogy would be a bilayer of two-dimensional worlds in 3D space, shown in figure 15. They are in contact everywhere, but oblivious of each other, despite their common interface. There is no freedom of motion,

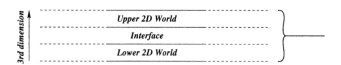

Figure 7.15: *Two flat worlds with a common interface.*

neither towards nor away from the interface. To cross the interface it is necessary to move into a third dimension, which is not an allowed operation in two dimensions. The only way to detect the presence of the interface is by interaction with the interfacial potential gradient. The gradient has little effect on massive entities, but influences microscopic particles dramatically. Likewise a massive three-dimensional world is everywhere in contact with the three-dimensional vacuum interface. In order to cross this interface it is necessary to proceed along a fourth dimension. The motion of a small particle, perturbed by the potential at the vacuum interface, is drastically influenced by undetected and hence ignored causes in the vacuum. Referred to known, observable influences only, the trajectory of the particle appears irrational, although mathematically readily tractable. Physics demands that the particle should follow a rational course, but observation shows that it moves like a wave. It will be argued that quantum theory is the mathematical account of this apparent wave motion.

The assumed element of PCT symmetry along the interface implies a closed one-sided space, wrapped around an interface that separates regions of opposed chirality. As known from the spontaneous resolution of racemates, chirality can be the driving force of phase separation. Objects with two-dimensional chirality can interconvert in three dimensions.

$$
\begin{array}{ccc}
A & & A \\
| & & | \\
B-X-C & \longleftrightarrow & C-X-B \\
| & & | \\
D & & D
\end{array}
$$

Although rotation and translation in the plane of the objects can never bring them into coincidence, rotation about an axis in the plane achieves the conversion. Such an inversion is also brought about by the Möbius twist, figure 8.

Three-dimensional chirality can likewise be resolved in four dimensions by transplantation in non-orientable projective space. This three-dimensional analog requires a four-dimensional twist that closes the three-dimensional universe onto itself and turns left-handed matter into right-handed anti-matter. Considered as a single universe in three-dimensional space, chirality is preserved throughout. However, the interface created by the curvature separates regions of space with opposite chiralities. The interface cannot be crossed in three-dimensional motion, but allows interaction between entities near the interface to give rise to the quantum effects.

A universe with this topology, viewed as a non-orientable manifold of period π, will be self-destructive and can obviously not be a serious model of the universe. The present proposal, however, is different: it pictures a universe based on the orientable double cover [231] of period 2π. The postulated interface, called the vacuum, is closed in four dimensions with period π, and corresponds to the relativistic hypersurface which is the locus of light signals and populated by bosons only. The normal to the surface which is oriented in space, is itself oriented in the fourth dimension u of Thierrin space. This condition reduces quantum behaviour to time-like fluctuations between regions of three-dimensional space.

7.4.1 The Quantum Potential

Quantum phenomena arise from the presence of the interfacial potential. To appreciate its effect, consider a sequence of classical particles in order of decreasing size. They become increasingly susceptible to the influence of the interfacial potential in the same order. At the interface, the potential gradient is very sharp and any assumption of constant potential, on the scale of the smallest particle, becomes totally untenable. This is the precise condition that differentiates between classical and non-classical behaviour.

The same argument differentiates between geometrical and wave optics. Geometrical optics follows the eikonal equation $(T(3.15))$ that assumes a constant index of refraction on the scale of the wavelength, and has the same form as the Hamilton-Jacobi equation of a classical particle $(T(3.5))$. In a field where potential energy suffers a large fractional change over a distance of the particle dimensions, the HJ, like the eikonal equation, no longer provides an adequate description of particle motion.

It is suggested that such conditions prevail near the vacuum interface. A sharp potential gradient, caused by the proximity of another phase, exists near the interface. Its perturbative effects increase with decreasing particle size, and this requires the HJ equation to be augmented for small particles. By analogy with the relationship between geometrical and wave optics the classical HJ equation is transformed into Schrödinger's equation (3.4), which is converted into (3.6) by substituting a wave function in polar form. It differs from the classical equation through the appearance of a term that defines the quantum potential.

The quantum potential can now be identified as a surface effect that exists close to any interface, in this case the vacuum interface. The non-local effects associated with the quantum potential also acquire a physical basis in the form of the vacuum interface, now recognized as the agent responsible for mediating the holistic entanglement of the universe. The causal interpretation of Bohmian mechanics finds immediate support in the postulate of a vacuum interface. There is no difference between classical and quantum entities, apart from size. Logically therefore, the quantum limit depends on the factor h/m that regulates the quantum potential.

7.4.2   Antimatter

It is not too difficult to identify anti-matter with the chiral opposite of matter and that the vacuum separates material and anti-material worlds. Transplantation along the double cover through a half period (π) can be described by a coordinate transformation $x_j \to -x_j$, $j = 1, 4$, and a gauge transformation $\Psi \to \Psi^* = \psi^* \exp(i\Phi_j x_j)$ $\big(T(4.45)\big)$. The 2π transplantation is an identity operation. The particle probability density $\langle\Psi|\Psi^*\rangle$ is gauge invariant but, because of the phase change, a transplantation of π turns particles into the conjugate, anti-particles of their former selves. Particle/anti-particle domains are not demarcated, but flow smoothly into each other, and an element of PCT symmetry appears along the vacuum interface. Exploration of the universe therefore never reveals the gradual change in chirality along the curvature in four dimensions, and all matter is perceived to be of the same chirality.

To understand why there is no mutual annihilation of matter and anti-matter that come into contact at the interface, it is noted that these phases are separated in the direction of the fourth coordinate u. By construction, points separated by the interface have their world lines (u) in opposite directions and therefore with perceived time flow inverted. This time difference prevents contact and hence annihilation across the interface.

7.4.3   Quantum Effects

The vacuum interface is the source of all quantum effects. Interaction with the interface causes particles to make excursions into time and bounce back with time-reversal and randomly perturbed space coordinates. Different from classical particles, quantum objects can suffer displacement in space without time advance. They can appear to be in more than one place at the same time, as in a two-slit experiment.

Awkward questions around the Dirac model of the vacuum (4.3) fall away. It is the quantum potential at the interface that prevents fermions from cascading down negative energy levels. The clear distinction between matter and anti-matter is described more directly by two linked equations such as a Schrödinger equation and its complex conjugate. The superluminal velocity component of an electron that manifests as Zitterbewegung (4.3.1), is no longer disallowed in the assumed Thierrin geometry. The trembling is caused by mixing, across the interface, of positive and negative energy states, exactly in line with the mathematical model proposed by Schrödinger [66].

Entities that move in the interface are achiral and massless. A virtual photon consists of a virtual particle/anti-particle pair. The vector bosons that mediate the weak interaction are massive and unlike photons, distinct from their anti-particles. The weak interaction therefore has reflection symmetry only across the vacuum interface and hence β-decay violates parity conservation.

Where the curvature of space is distorted by the presence of large masses, such as neutron stars, quasars or black holes, the interfacial potential gradient is enhanced and quantum effects are predicted to become more pronounced in the vicinity of high-density material. The result is an increased value of h and more pronounced quantum uncertainty. More massive entities show quantum behaviour and in the limit of infinite gradient, the interface is ruptured and uncertainty becomes total, $\Delta E \cdot \Delta t \geq h$. This situation corresponds to a black hole. Increased h also produces radiation of constant-energy quanta at lower frequency, amounting to an intrinsic red shift, as observed with quasars. In general, seepage across the interface contributes to cosmic radiation and the microwave background. The black-body spectrum of the latter is an inevitable consequence of the proposed closed topology of space.

Another consequence of the general curvature of world space is that distant radiation sources are separated from an observer in both space and time [232]. During transit, the photon moves towards an observer which is ahead in time, and therefore appears to lag as if the source was receding. The observed red shift, created by the time difference, will be a function of separation in space and proportional to the time interval, Δt. The red shift and Hubble's proportionality constant are then defined by

$$z = \Delta\lambda/\lambda \;=\; Hr/c$$
$$= \left[\frac{1}{t} - \frac{1}{t+\Delta t}\right]\frac{r}{c}$$
$$\text{Since} \quad r = tc, \quad z \;=\; 1 - \frac{t}{t+\Delta t}$$

As $\Delta t \to \infty$, $H \to 1/t_o = c/r_o$, where r_o is interpreted as a Hubble radius of the universe.

Earlier speculations about the effect of the curvature of space on elemental synthesis and the stability of nuclides (2.4.1) are consistent with the interface model. The absolute curvature of the closed double cover of projective space, and the Hubble radius of the universe, together define the golden mean as a universal shape factor [233], characteristic of intergalactic space. This factor regulates the proton:neutron ratio of stable nuclides and the detail of elemental periodicity. The self-similarity between material structures at different levels of size, such as elementary particles, atomic nuclei, chemical

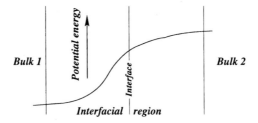

Figure 7.16: *Variation of chemical potential in the vicinity of an interface between different phases.*

atoms, biological organisms, solar systems and spiral galaxies, depends on the same shape factor. As one example, the presumed standing-wave structure of the electron (figure 4.1) follows the same boundary curve (figure 7.14) that, when shrunk away, produces the projective world space assumed here.

7.5   Chemical Effects

Quantum phenomena at the vacuum interface have been postulated in analogy with known effects at physico-chemical interfaces. To be consistent, special properties of the latter are therefore implied. A physical interface is the boundary surface that separates two phases in contact. These phases could be two solid phases, two liquid phases, solid-liquid, solid-gas or liquid-gas phases. What they all have in common is a potential difference between the two bulk phases. In order to establish equilibrium at the interface it is necessary that rearrangement occurs on both sides of the interface over a narrow region. Chemical effects within the interfacial zone are unique and responsible for the importance of surfaces in chemical systems. At the most fundamental level the special properties of surfaces relate to the difference between isolated elementary entities and the same entities in a bulk medium, or condensed phase.

Atomic and sub-atomic particles behave fundamentally different from macroscopic objects because of quantum effects. The more closely an atom is confined the more classical its behaviour. (Compare 5.2.1). Mathematically, the boundary condition on the particle wave function $\psi \to 0$ as $r \to \infty$, is replaced by $\lim_{r \to r_o} \psi = 0$, where $r_o << \infty$. It means that the influence of the free particle has a much longer reach through its wave function than a particle confined to a bulk phase. Wave-mechanically, the wavelength of the particle increases and approaches infinity for a completely localized, or classical particle. Electrons and atoms in condensed phases, where their motion is severely restricted, behave more classically and they are in equilibrium with their environment.

The physics of condensed phases is commonly formulated as of infinite extent. However, solid and liquid objects in the laboratory are of finite size and terminate discontinuously in a surface (in vacuum) or an interface, under all other conditions. Atoms or molecules at the surface or interface of the condensed object find themselves in a completely different environment, compared to those in the interior of the body. They are less confined in at least one direction, which means that the wave function looks different in this direction - it is less classical. It is implied that surface or interfacial species show more quantum-mechanical behaviour, compared to the bulk. This is the basic reason for the special properties of surfaces and the origin of all interfacial phenomena. Surface chemistry should therefore be formulated strictly in terms of quantum theory, but this has never been attempted. In its present state of development it still is an empirical science, although many physico-chemical concepts are introduced to rationalize the behaviour of interfaces.

It becomes obvious why surface effects were first observed and studied in the colloidal state. In this case one deals with particles that are so small that the number of atoms or molecules in the surface represents a substantial fraction of all the material that makes up the colloidal system. Colloidal particles are defined as having at least one dimension in the range of 1 micron to 1 nm. In modern chemistry however, particles in the nanometer range are given a special name, *nanoparticles*. The reason is that in this range several completely unpredictable phenomena start to appear [234]. The reason is obvious: these particles behave largely quantum-mechanically and therefore completely differently from ordinary colloidal particles. In many instances unusual mechanical, acoustical, electrical and optical properties are associated with nanomaterials, which have also been mentioned in connection with controversial issues such as room-temperature superconductivity and cold fusion.

7.5.1 Interfacial Equilibrium

The Boltzmann distribution (T 6.2) describes how a set of identical particles distribute themselves in a potential-energy field. At equilibrium the mole fractions of particles X_1 and X_2 at two levels in the field, are given by

$$X_1 = X_2 \exp\left[-\left(\mu_1^i - \mu_2^i\right)/kT\right]$$

where the μ_n^i are potential energies of species i. This equation can be written as

$$\mu_1^i + kT \ln X_1 = \mu_2^i + kT \ln X_2$$

If there are many regions or states in a system, each at a different potential μ_n^i, the equilibrium condition can be generalized into

$$\mu_n^i + kT \ln X_n = \text{constant, for } n = 1, 2, 3 \ldots$$
$$= \mu \qquad (7.4)$$

In other words, there will be a flow of molecules between all the different states until (4) is satisfied, *i.e.* equilibrium is established when the value of $\mu_i + kT \ln X_i$ is uniform throughout. The quantity μ is the chemical potential (T 8.2.2) and it specifies the total free energy per molecule. It includes interaction and thermal energies. The factor $k \ln X_n$ gives the entropy of confining the molecules.

At the interface in a two-phase system the difference in chemical potential of each component is due to different intermolecular interactions in the two phases. If one of the phases ($n = 1$, say) is a pure substance, $\ln X_1 = 0$,

$$\mu_1^i = \mu_2^i + kT \ln X_2$$

and

$$X_2 = X_1 \exp\left(-\Delta\mu^i / kT\right)$$

For example, μ_2^i could be the energy of molecules in solution relative to their energy in the solid, μ_1^i. An amount of X_2 molecules will dissolve in order to establish an interfacial equilibrium.

The effect of an interface on bulk phases is illustrated by the structures generated in a liquid phase in contact with an inert solid surface [235]. Although molecules in the liquid state move around more freely than in solids, their motion is still restricted by intermolecular interaction. The way they position themselves with respect to each other depends on their shape and the mean intermolecular separation. The tendency to pack in an ordered array persists in liquids. Next-nearest neighbours order around the first ordered cluster, but with a lower degree of correlation. Eventually, at large distances there will be no correlation with respect to the reference molecule, which may be considered fixed to the solid surface. As a working model it is assumed that the entire solid surface is covered with molecules of the liquid phase immobilized in the surface. Their combined effect on the bulk liquid will resemble the radial distribution function of a single molecule, as a fluctuating density field perpendicular to the surface. A liquid medium near the surface will therefore not be a structureless continuum. The density

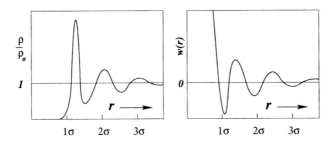

Figure 7.17: *Left: Fluctuating density in a liquid phase near a solid surface, relative to the bulk density ρ_o. The first maximum occurs at the hard-sphere contact distance from the surface, at $r = \sigma$ the molecular diameter. Right: Pair potentials for two molecules in the liquid calculated from* $\rho(r) = \rho_o \exp[-w/kT]$

near the surface fluctuates as shown in figure 17, until it reaches bulk density several molecular diameters (σ) away from the surface. Molecular interaction in the surface region is governed by pair-wise potentials that follow the fluctuation in density and appears as an oscillatory force, apparently seated in the interface.

The interface between two liquid phases will differ from this construct in detail only. The postulated effects of a potential field that changes appreciably over the dimensions of interacting particles near the surface remain valid. In the case of the vacuum interface it is the quantum-potential field that causes the surface effects.

7.5.2 Non-local Interaction

Quantum interaction is non-local. In small molecules all interactions, mediated by quantum-mechanical exchange of electrons, are of this type and with a range that depends on the quantum potential. The largest globular proteins have diameters in the nanometer range. Although the largest DNA molecule can reach 1 meter in length, volumewise this corresponds to a sphere of diameter in the micron range. It seems reasonable to conclude that the chemistry of all molecules, including biological macromolecules, depends on non-local effects.

An ancient puzzle, immediately addressed by the previous conclusion concerns intramolecular rearrangements that appear unlikely on stereochemical grounds. The most daunting of these is the uncoiling of double-stranded DNA and the related folding of proteins.

A meter-long double helix of DNA compacted into a cell nucleus of micron size is observed to unwind in a matter of minutes. There is no mechanical model that can possibly account for this operation that requires the tangled double helix to rotate smoothly at a thousand rpm about its long axis, without aggravating the tangle. This unwinding is related to the action of flagella, the organelles of locomotion in many plants and animals. Flagellum, a type of bacterium, rotates at over 10 000 rpm, driven by proton flow under an electrochemical potential difference across a membrane. The diameter of the bearing is about 20 nm with an estimated clearing of about 1 nm. The precise mechanism is not known. At this stage the details of non-local interaction are completely obscure, but no other reasonable explanation of these effects exists. Intramolecular non-local effects could take the form of superconductivity, superfluidity or long-range tunnelling. The physics of intramolecular space is simply unexplored.

Protein folding is evidently not going to find a reasonable explanation in terms of classical stereochemistry. One of the more remarkable features is the effect of the environment on the tertiary structure of a protein. Not only does a protein when moved between media of different polarity become denatured - it could, under carefully controlled conditions invert its pattern of folding reversibly when cycled between such media. It is as if hydrophobic regions on the one hand and hydrophilic regions on the other, are in long-range communication to decide on the optimal self-assembly to neutralize amphiphilic imbalances. All efforts to model protein folding by energy minimization, using pairwise potentials have been singularly unsuccessful. As with DNA, it is convenient for the time being, to hide behind collective ignorance of non-local effects and only stimulate interest in new possibilities of how to approach these problems in future.

Exploration of intramolecular non-local effects could be the beginning of more far-reaching studies. Neural receptors with the ability to exchange information via the quantum-potential field in the vacuum interface, could be another level of quantum object that might eventually explain para-psychic phenomena.

The special effects associated with materials made up of nanosized particles are also due to non-local interactions. Such effects were first noticed in the field of micro-electronics during efforts to decrease the size of devices to quasi one-dimensional or zero-dimensional structures [236]. One possibility for obtaining zero-dimensional structures is the inclusion of spherical semiconductor particles in a transparent dielectric medium. Such isolated microcrystals, typically of nanometer size are known as quantum dots. Best-known examples include particles of CdS and CdSe isolated in silicate glasses.

The production of nanocrystals stimulated fresh research into the understanding of nucleation and growth of crystals. The number of nuclei per embryo was found to strongly depend on the degree of supersaturation. The critical cluster radius at which the surface free energy has a maximum, follows from the Thomson-Gibbs equation [237],

$$r^* = \frac{2\gamma v_c}{k_B T \ln(p/p_o)}$$

This equation is identical to the classical Kelvin equation (5.13) and the main conclusions are consistent with the nucleation model of section 5.10.1. As pointed out there, nanosized clusters, with the hidden symmetry of bulk solid, exist in the critical fluid, available to be captured as quantum dots in a glassy matrix. At this stage the crystalline structure of the bulk material has already been developed and the effects of quantum confinement can be observed.

Alternative chemical strategies for the synthesis of nanomaterials, like the organization in the critical region, also rely on self-assembly of suitable monomers. Cyclic oligopeptides may, for instance, be synthesized by linking eight amino acid residues by covalent bonds. The resulting macrocycles self-assemble into nanoscaled tube-like arrays.

Single-wall carbon nanotubes are new types of nanomaterial, the study of which generates about five research papers [234] from around the world, each day. An important feature of these structures is that the aromatic rings of the folded graphite sheet that constitutes the tube, are no longer planar. This feature represents a new challenge for accepted theories of π-bonding.

Bibliography

[1] P.A.M. Dirac, *Proc. Roy. Soc. (Lond.)*, 1929, **A123**, 714.

[2] H. Primas, *Chemistry, Quantum Mechanics and Reductionism*, 2nd ed., 1983, Springer-Verlag, Berlin.

[3] A. Einstein, B. Podolsky and N. Rosen, *Phys. Rev.*, 1935, **47**, 777.

[4] D. Bohm, *Phys. Rev.*,1952, **85**, 166;180.

[5] J.S. Bell, *Found. of Phys.*, 1982, **12**, 989.

[6] D.W. Belousek, *Stud. Hist. Phil. Mod. Phys.*, 1996, **27**, 437.

[7] J.C.A. Boeyens, *The Theories of Chemistry*, 2003, Elsevier, Amsterdam.

[8] J.C.A. Boeyens, *Crystal Engineering*, 2001, **4**, 61.

[9] J.C.A. Boeyens, *Crystal Engineering*, 2003, **6**, 167.

[10] J.C.A. Boeyens, *Trans. Roy. Soc. S. Afr.*, 1999, **54**, 323.

[11] J.C.A. Boeyens, *S. Afr. J. Chem.*, 2000, **53**, 49.

[12] P.A.M. Dirac in [13].

[13] B.Gruber and R.S. Millman, eds., *Symmetries in Science*, 1980, Plenum, N.Y.

[14] J. Rosen, *A Symmetry Primer for Scientists,* 1983, Wiley,

[15] H. Goldstein, *Classical Mechanics*, 2nd ed., 1980, Addison-Wesley, Reading.

[16] H.D. Fegan, *Introduction to Compact Lie Groups*, 1991, World Scientific, Singapore.

[17] W. Greiner and B. Müller, *Quantum Mechanics. Symmetries*, 2nd. ed., 1994, Springer, Heidelberg.

[18] J. Goldstone, *Il Nuovo Cimento*, 1961, **19**, 154.

[19] F. Halzen and A.D. Martin, *Quarks and Leptons: An Introductory Course in Modern Particle Physics*, 1984, Wiley, N.Y.

[20] A-E. B. de Chancourtois, (in translation), *Nature*, 1889, **41**, 186.

[21] J.C.A. Boeyens, *J. Radioanal. Nucl. Chem.*, 2003, **257**, 33.

[22] J.E. Reynolds, *J. Chem. Soc.*, 1902, **81**, 612.

[23] W.N. Cottingham and D.A. Greenwood, *An introduction to nuclear physics*, 1986, Cambridge University Press, Cambridge.

[24] J.C.A. Boeyens, *J. Chem. Soc., Faraday Trans.*, 1994, **90**, 3377.

[25] S. Goldman and C. Joslin, *J. Phys. Chem.* 1992, **96**, 6021.

[26] R. Adler, M. Bazin and M. Schiffer, *Introduction to General Relativity,* 1965, McGraw-Hill, N.Y.

[27] A. Beck, M.N. Bleicher and D.W. Crowe, *Excursions into Mathematics,* 2000, AK Peters.

[28] J.C.A. Boeyens, *Specul. Science and Tech.* 1992, **15**, 192.

[29] M. Friedman, *Foundation of Space-Time Theories,* 1983, University Press, Prineton, NJ.

[30] E. Schrödinger. *Ann. Phys.,* 1926, **79**,489.

[31] E. Madelung, *Z. Phys.,* 1926, **40**, 322.

[32] L. de Broglie, *C.R. Acad. Sci. Paris,* 1926, **183**, 447; 1927, **184**, 273 ; **185**, 380.

[33] D. Home, *Conceptual Foundations of Quantum Physics,* 1997, Plenum, N.Y.

[34] D. Bohm and B.J. Hiley, *The undivided universe. An ontological interpretation of quantum theory,* Routledge, London, 1993.

[35] P.R. Holland, *The quantum theory of motion,* University Press, Cambridge, 1993.

[36] T. Takabayasi, *Progr. Theoret. Phys. (Japan),* 1952, **8**, 143.

[37] D. Bohm and J.P. Vigier, *Phys. Rev.,* 1954, **96**, 208.

[38] L. de Broglie, *Compt. rend.,* 1953, **235**, 1345, 1372.

[39] D. Bohm, *Quantum Theory,* 1951, Prentice-Hall, Englewood Cliffs, N.J.

[40] J.S. Bell, *Physics,* 1964, **1**, 195.

[41] J. Clauser, M. Horn, A. Shimony and R. Holt, *Phys. Rev. Lett.,* 1969, **26**, 880.

[42] A. Einstein in M. Born(ed.) *Albert Einstein Max Born Briefwechsel 1916 – 1955,* Wiener Verlag, Himberg, 1991.

[43] J.F. Clauser and A. Shimony, *Rep. Prog. Phys.* 1978, **41**, 1881.

[44] A. Aspect, J. Dalibard and G. Roger, *Phys. Rev. Lett.,* 1982, **49**, 1804.

[45] J.C.A. Boeyens, *S. Afr. J. Chem.,* 1986, **82**, 361.

[46] J.C. Smuts, *Holism and Evolution,* 1926, Macmillan, London.

[47] A. Einstein, *Dialectica,* 1948, **2**, 320.

[48] N. Huggett, *Am. J. Phys.* 1997, **65**, 1135.

[49] J.D. Gervey, *Introduction to Modern Physics,* Academic Press, N.Y., 1971.

[50] J-M. Lévy-Leblond, *Am. J. Phys.* 1976, **44**, 719.

[51] S. Glasstone, *Textbook of Physical Chemistry,* 1940, Macmillan, London.

[52] G.J. Stoney, *Proc. Dublin Soc.,* 1881, **3**, 51.

[53] G.N. Lewis, *J. Am. Chem. Soc.,* 1916, **38**, 762-785.

[54] R.W. Gurney, *Proc. Roy. Soc. (London),*1932, **134A**, 137.

[55] J.D. Jackson, 1975. *Classical Electrodynamics*, 2nd ed., New York,Wiley.

[56] H. Poincaré, *C.R. Acad. Sci.*, 1905, **140**, 1504.

[57] P.A.M. Dirac, *Proc. Roy. Soc. (London)*, 1938, **A167**: 148.

[58] E.J. Konopinsky, 1981.*Electromagnetic Fields and Relativistic Particles*, N.Y. McGraw-Hill.

[59] P.A.M. Dirac, *Proceedings of the Royal Society (London)*, 1951, **A209**: 291-296.

[60] J. Horgan, *New Scientist*, 1993, 27 February: 38.

[61] J.R. Oppenheimer, *Phys. Rev.*, 1930, **35**, 461.

[62] P.A.M.Dirac, *Proc. Roy. Soc. (London)*, 1928, **A117**, 351

[63] S. Saunders, in [64]

[64] S. Saunders and H.R. Brown, *The Philosophy of Vacuum*. 1991, Clarendon Press, Oxford.

[65] P.A.M. Dirac, *Proc. Roy. Soc. (London)*, 1928, **A117**: 610.

[66] E. Schrödinger, *Sitzungsberichte der Preusssischen Akademie der Wissenschaften*, 1931, 63.

[67] E. Schrödinger, *Sitzungsberichte der Preussischen Akademie der Wissenschaften*, 1930, 418.

[68] Winterberg, *Zeitschrift für Naturforschung*, 1991, **46a**, 746, and references therein.

[69] F.M. Meno, *Physics Essays*, 1991, **4**, 1.

[70] A. Rüger, *Hist. Stud. Phil. Sci.*, 1988, **18**, 377.

[71] A. S. Eddington, *New Theories in Physics*, Paris, 1939.

[72] E. Schrödinger, *Nature*, 1937, **140**, 742.

[73] R.C. Jennison, and A.J. Drinkwater, *Journal of Physics A*, 1977, **10**, 167.

[74] A. Einstein, *Ann. Physik. Leipzig*, 1905, **17**, 891.

[75] R.C. Jennison, *Journal of Physics A*, 1978, **11**, 1525.

[76] A. Sommerfeld, and H. Welker, *Ann. Physik*, 1938, **32**, 56.

[77] H. Margenau, and G.M. Murphy, *The Mathematics of Physics and Chemistry*, 2nd ed., 1956, Van Nostrand, Princeton.

[78] J.G. Cramer, *Phys. Rev. D*, 1980, **22**, 362.

[79] C. Elbaz, *Compt. Rend. Acad. Sci. II*, 1983, **297**, 455.

[80] C. Elbaz, *Physics Letters A*, 1985, **109**, 7.

[81] C. Elbaz, *Physics Letters A*, 1986, **114**, 445.

[82] C. Elbaz, *Physics Letters A*, 1987, **123**, 205.

[83] C. Elbaz, 1988. *Physics Letters A*, 1988, **127**, 308.

[84] M. Wolff, *Galileon Electrodynamics*, 1995, **6**, 83.

[85] M. Wolff, *Frontier Perspectives*, 1997, 17 pages. (Preprint).

[86] M. Molski, *Advances in Imaging and Electron Physics*, 1998, **101**, 143.

[87] H.C. Corben, *Lettere al Nuovo Cimento*, 1977, **20**, 645.

[88] O. Klein, *Nature*, 1926, **118**, 516.

[89] Th. Kaluza, *Sitzungsberichte der Preussischen Akademie der Wissenschaften*, 1921, 966.

[90] O. Klein, *Z. Physik*, 1926, **37**, 895.

[91] M. Molski, *Physics Essays*, 1997, **10**, 18.

[92] Y. Thirdy, *C.R. Acad. Sci.*, 1948, **226**, 516.

[93] R. Weingard, in [64]

[94] E.P. Battey-Pratt, and T.J. Racey, *Int. J Theor. Phys.*, 1980, **19**, 437.

[95] C.W. Misner, K. Thorne, and J.A. Wheeler, *Gravitation*, 1973, San Francisco. Freeman, Chapter 41. Spinors.

[96] H.C. Corben, and P. Stehle, *Classical Mechanics*, 1950, New York.Wiley.

[97] J.C.A. Boeyens, *J. Chem. Ed.*, 1995, **72**, 412.

[98] M. Carmeli, *Lettere al Nuovo Cimento*, 1985, **42**, 67.

[99] D.J. Griffiths, *Introduction to Electrodynamics*, 1981, New Jersey. Prentice-Hall. p. 287.

[100] R.H. Romer, *Am. J. Phys.*, 1967, **35**, 445.

[101] L. de Broglie, *Non-Linear Wave Mechanics*, 1960, Translation from the French by A.J. Knodel and J.C. Miller, Amsterdam. Elsevier.

[102] D. Bohm, R. Schiller, and J. Tiomno, *Nuovo Cimento, Supplemento*, 1955, **1**, 48.

[103] H. Tetrode, *Z. Phys.*, 1922, **10**, 317.

[104] G.N. Lewis, *Proc. Nat. Acad. Sci. U.S.*, 1926, **12**, 22.

[105] I.G. Naan, *The symmetrical universe*, (Translation). *Academy of Sciences of the Republic of Estonia. Publication of Tartu Astronomical Observatory*, 1964, **34**, # 6, 13 pages.

[106] J.A. Wheeler, and R.P. Feynman, *Rev. Mod. Phys.*, 1945, **17**, 157.

[107] J.G. Cramer, *Rev. Mod. Phys.*, 1986, **58**, 647.

[108] J.C.A. Boeyens and J. du Toit, *Electr. J. Theoret. Chem.*, 1997, **2**, 296.

[109] A. Michels, J. de Boer, and A. Bijl, *Physica*, 1937, **4**, 981.

[110] S.R. de Groot and C.A. ten Seldam, *Physica*, 1946, **12**, 669.

[111] P. Plath, Dissertation, Technische Universität Berlin, 1972.

[112] P.O. Fröman, S. Yngne and N. Fröman, *J. Math. Phys.*, 1987, **28**, 1813.

[113] Y.P. Varshni, *J. Phys. B*, 1998, **32**, 2849.

[114] J.C.A. Boeyens, *Electr. J. Theor. Chem.*, 1995, **1**, 38.

[115] J.C.A. Boeyens, *S. Afr. J. Chem.*, 1980, **33**, 63.

[116] M. Born and R. Oppenheimer, *Ann. Phys.*, 1927, **84**, 457.

[117] L.C. Pauling, *The Nature of the Chemical Bond*, 3rd ed., Cornell UP, 1960.

[118] R.S. Mulliken, *J. Chem. Phys.*, 1934, **2**, 782.

[119] M.C. Day and J. Selbin, *Theoretical Inorganic Chemistry*, 2nd ed., 1969, Van Nostrand.

[120] J. Hinze and H.H. Jaffe, *J. Am. Chem. Soc.*, 1962, **85**, 148.

[121] A.L. Allred and E.G. Rochow, *J. Inorg. Nucl. Chem.*, 1958, **5**, 264.

[122] R.G. Parr and R.G. Pearson, *J. Am. Chem. Soc.*, 1983, **105**, 7512.

[123] R.P Iczkowski and J.L. Margrave, *J. Am Chem. Soc.*, 1961, **83**, 3547.

[124] R.G. Parr and R.G. Pearson, *J. Am. Chem. Soc.*, 1983, **105**, 7512.

[125] F. Herman and S. Skillman, *Atomic Structure Calculations*, Prentice-Hall, NJ, 1963.

[126] J.K. Nagle, *J. Am. Chem. Soc.*,1990, **112,** 4741.

[127] J. du Toit, *Chemical Applications of Ionization Radii*, Ph.D. thesis, University of the Witwatersrand, 1997.

[128] L.C. Allen, *J. Am. Chem. Soc.*, 1992, **114,** 1510.

[129] L.C. Allen and E.T. Knight, *J. Mol. Struct. (Theochem)*, 1992, **261,** 313.

[130] L.C. Allen, *J. Am. Chem. Soc.*, 1989, **111,** 9003.

[131] A.S. Eddington, *The Nature of the Physical World*, Cambridge UP, 1928.

[132] W.T. Grandy, *Introduction to Electrodynamics and Radiation*, Academic Press, NY, 1970.

[133] A. Amann in [134]

[134] W. Gans and J.C.A. Boeyens (eds), *Intermolecular Interactions*, Plenum, New York, 1998.

[135] J.C.A. Boeyens, *J. S. Afr. Chem. Inst.*, 1973, **26**, 94.

[136] J.C.A. Boeyens in [134]

[137] R.T. Sanderson, *Chemical bonds and bond energy*, Academic Press, N.Y., 1971

[138] H. Eyring, J. Walter and G.E. Kimball, *Quantum Chemistry*, 1944, Wiley, New York.

[139] J.C. Slater, *Quantum Theory of Molecules and Solids*, Vol. 1, 1963, McGraw-Hill, New York.

[140] J. Edwards, *A Treatise on the Integral Calculus*, Vol. 2, 1922, p. 130, Macmillan, London.

[141] T.L. Cottrell and L.E. Sutton, *Proc. Roy. Soc. (London)*, 1951, **A207**, 49.

[142] J.C.A. Boeyens, *S. Afr J. Chem.*, 1980, **33**, 14.

[143] I.D Brown, *Chem. Soc. Revs.*, 1978, **7**, 359.

[144] J.C.A. Boeyens, *J. Cryst. Spectr. Res.*, 1982, **12**, 245.

[145] J.C.A. Boeyens and D.J. Ledwidge, *Inorg. Chem.*, 1983, **22**, 3587.

[146] P.M. Morse, *Phys. Rev.*, 1929, **34**, 57.

[147] T. Shimanouchi in D. Henderson (ed.) *Physical Chemistry, An Advanced Treatise*, Vol. IV, 1970, Academic Press, New York.

[148] A.Y. Meyer, *J. Mol. Struct. (Theochem)*, 1988, **179**, 83.

[149] A. Einstein, *Ann. Physik*, 1916, **49**, 769.

[150] M. Carmeli, *Group Theory and General Relativity*, 2000, Imperial College Press, London.

[151] J.W. Mullin, *Crystallization*, 2nd ed., 1972, Butterworths, London.M

[152] J.G. Kirkwood, in [153].

[153] R. Smoluckowski, J.E. Mayer and W.A Weyl, *Phase Transformations in Solids*, 1951, Wiley, N.Y.

[154] R. Smoluckowski, in [153].

[155] V.A. Garten and R.B. Head, *Phil. Mag.*, 1963, **8**, 1793; 1966, **14**, 1243.

[156] M. von Laue, *Z. Kristallogr.*, 1943, **105**, 124.

[157] C. Kittel, *Introduction to Solid State Physics*, 5th ed., 1976, Wiley, N.Y.

[158] I.E. Segal in [13].

[159] W.G. Dixon, *Special Relativity*, 1978, University Press, Cambridge.

[160] I. Prigogine, *Int. J. Theor. Phys.*, 1989, **28**, 927.

[161] R.G. Woolley, *Adv. Phys.*, 1976, **45**, 29.

[162] J.C.A. Boeyens, *Structure and Bonding*, 1985, **63**, 65.

[163] A. Amann, *S Afr. J. Chem.*, 1992, **45**, 29.

[164] B.T. Sutcliffe, *J. Mol. Struct. (Theochem)*, 1992, **29**, 259.

[165] P. Claverie, and S. Diner, *Israel J. Chem.*, 1980, **19**, 54.

[166] L. de Broglie, *Nature*, 1926, **118**, 441.

[167] N. Bohr, *Dialectica*, 1948, **2**, 312.

[168] J.F. Ogilvie in E.S. Kryachko and J.L. Calais (eds), *Conceptual Trends in Quantum Chemistry*, 1994, Kluwer, Dordrecht.

[169] L. Pauling, *J. Am. Chem. Soc.* 1931, **53**, 1367, 3225.

[170] R.J. Gillespie, *J. Chem. Ed.*, 1970, **47**, 18.

[171] R.J. Gillespie, *Molecular geometry*, 1972, Van Nostrand, London.

[172] R.J. Gillespie and R.S. Nyholm, *Quart. Revs. Chem. Soc.*, 1957, **11**, 339.

[173] R.F.W. Bader, *Atoms in Molecules*, 1994, Clarendon Press, Oxford.

[174] J.C.A. Boeyens *S. Afr. J. Chem.*, 1980, **33,** 66.

[175] S.S. Batsanov, *Zh. Strukt. Khim.*, 1978, **19,** 958.

[176] J.C.A. Boeyens in [177].

[177] W.Gans, A. Amann and J.C.A. Boeyens (eds.) *Fundamental Principles of Molecular Modeling*, Plenum, N.Y., 1996.

[178] R.G. Woolley, *J. Am. Chem. Soc.*, 1978, **100**, 1073.

[179] T. Koritsanszky in [177].

[180] T.S. Koritsanszky and P. Coppens, *Chem. Revs.*, 2001, **101**, 1583.

[181] J.D. Dunitz and P. Seiler, *J. Am. Chem. Soc.*, 1983, **105**, 7056.

[182] W.H.E. Schwarz, K. Ruedenberg and L. Mensching, *J. Am. Chem. Soc.*, 1989, **111**, 6926.

[183] T. Berlin, *J. Chem. Phys.*, 1951, **105**, 7056.

[184] R. Daudel in P. Becker (ed.) *Electron and Magnetization Densities in Molecules and Crystals*, NATO ASI, Series B-Physics, 1980, **40**.

[185] P. Pfeifer, *Chiral Molecules - a Superselection Rule Induced by the Radiation Field.*, 1980, Dissertation ETH 6551, Swiss Federal Institute of Technology.

[186] F. Hund, *Z. Phys.*, 1927, **43**, 805.

[187] L. Rosenfeld, *Z. Phys.*, 1929, **52**, 161.

[188] M. Born and P. Jordan, *Elementare Quantenmechanik*, Springer, Berlin, 1930.

[189] C.K. Jørgenson, *Theor. Chem. Acta*, 1974, **34**, 189.

[190] E.B. Davies, *Comm. Math. Phys.*, 1979, **64**, 191.

[191] R.G. Woolley, *Isr. J. Chem.*, 1980, **19**, 30.

[192] S.F.A. Kettle, *Symmetry and Structure*, Wiley, N.Y., 1995.

[193] B.T. Sutcliffe in [177]

[194] E. Clementi and D.L. Raimondi, *J. Chem. Phys.* 38 (1963) 2686.

[195] A. Amann in [177].

[196] J.C.A. Boeyens, F.A. Cotton and S. Han, *Inorg. Chem.*, 1985, **24**, 1750.

[197] A.H. Morrish, *The Physical Principles of Magnetism*, Wiley, New York, 1965.

[198] J.C.A. Boeyens and S.M. Dobson, in I. Bernal (ed.) *Stereochemistry of Organometallic and Inorganic Compounds*. Volume 2, 1987, Elsevier, Amsterdam.

[199] W. Klyne and V. Prelog, *Experientia*, 1960, **16**, 521.

[200] J.E. Kilpatrick, K.S. Pitzer and R. Spitzer, *J. Am. Chem. Soc.*, 1947, **69**, 2483.

[201] J.B. Hendrickson, *J. Am. Chem. Soc.*, 1961, **83**, 4537; 1964, **86**, 4854; 1967, **89**, 7036.

[202] H.J. Geise, W.J. Adams, and L.S. Bartlett, *Tetrahedron*, 1969, **25**, 3045.

[203] H.M. Pickett and H.L. Strauss, *J. Am. Chem. Soc.*, 1970, **92**, 7281; *J. Chem. Phys.*, 1971, **55**, 324.

[204] J. Dale, *Acta. Chem. Scand.*, 1973, **27**, 1115.

[205] D. Cremer and J.A. Pople, *J. Am. Chem. Soc.*, 1975, **97**, 1354.

[206] D. Cremer, *Israel J. Chem.*, 1980, **20**, 12.

[207] J.C.A. Boeyens and D.G. Evans, *Acta Cryst. B*, 1989, **45**, 577.

[208] D. Evans and J.C.A. Boeyens, *Acta Cryst. B*, 1989, **45**, 581.

[209] A.E.H. Love, *A Treatise on the Mathematical Theory of Elasticity*, Section 293b, 1927, Cambridge University Press, London.

[210] P. Comba and T.W. Hambley, *Molecular Modeling of Inorganic Compounds*, 2nd ed., 2001, Wiley-VCH, Weinheim.

[211] F.H. Allen, in [177]

[212] A.Y. Meyer, *J. Chem. Soc. Perkin Trans.*, 1986, 1567.

[213] P.G. Mezey, *Int. Rev. Phys. Chem.*, 1997, **16**, 361.

[214] A.S. Mitchell and M.A. Spackman, *J. Comput. Chem.*, 2000, **21**, 933.

[215] P. Coppens, *X-ray Charge Densities and Chemical Bonding*, 1997, Oxford University Press, New York.

[216] V. Prelog, *J. Chem. Soc.*, 1950, 420.

[217] S. Shaik, P. Maitre, G. Sini and P.C. Hiberty, *J. Am. Chem. Soc.*, 1992, *114*, 7861.

[218] C.N. Yang in C.W. Kilmister (ed.) *Schrödinger. Centenary celebration of a polymath*, 1987, University Press, Cambridge.

[219] E. Schrödinger, *Z. Phys.*, 1922, **12**, 13.

[220] F. London, *Z. Phys.*, 1927, **42**, 375.

[221] E. Schrödinger, *Comp. Rend., Paris*, 1933, **1**, 581.

[222] A. Einstein and W. Mayer, *Sitz. Ber. Preuss. Akad.*, 1931, 541.

[223] P.S. Wesson, *Space, Time, Matter: Modern Kaluza-Klein Theory*, 1999, World Scientific, Singapore.

[224] G. Thierrin, *J. of Theoretics*, Vol 3-4, 2001.

[225] H.S.M. Coxeter, *Introduction to Geometry*, 2nd ed., 1989, Wiley, USA.

[226] H.G. Flegg, *From Geometry to Topology*, 1974, Dover Publications, Mineola, N.Y.

[227] J. Fauvel, R. Flood and R. Wilson, *Möbius and his band*, 1993, University Press, Oxford.

[228] J.V. Narlikar, *An Introduction to Cosmology*, 3rd ed., 2002, University Press, Cambridge.

[229] B. Roukema, *Monthly Notices Roy. Astron. Soc.*, 1996, **283**, 1147.

[230] P.A.M. Dirac, *Nature*, 1951, **168**, 906.

[231] M.P. de Carmo, *Differential Geometry of Curves and Surfaces*, 1976, Prentice-Hall, New Jersey.

[232] J.C.A. Boeyens, *S. Afr. J. Sci.*, 1995, **91**, 220.

[233] M. Livio, *The Golden Ratio*, 2002, Review, London.

[234] B. Bhushan (ed.) *Spinger Handbook of Nanotechnology*, 2004, Springer, Berlin.

[235] J.N. Israelachvili, *Intermolecular and Surface Forces*, 2nd ed., 1992, Academic Press, London.

[236] U. Woggon, *Optical Properties of Semiconductor Quantum Dots*, 1997, Springer, Berlin.

[237] B. Mutaftschiev in D.T.J. Hurle (ed.) *Handbook of Crystal Growth*, Vol. 1, 1993, Elsevier.

Index

266